CHEMICAL ENGINEERING METHODS AND TECHNOLOGY

PREPARATION OF THIN FILM PD MEMBRANES FOR H2 SEPARATION FROM SYNTHESIS GAS AND DETAILED DESIGN OF A PERMEABILITY TESTING UNIT

CHEMICAL ENGINEERING METHODS AND TECHNOLOGY

Handbook of Membrane Research: Properties, Performance and Applications
Stephan V. Gorley
(Editor)
2009. ISBN: 978-1-60741-638-8

Non-Ionic Surfactants
Pierce L. Wendt and Demario S. Hoysted
(Editors)
2009. ISBN: 978-1-60741-434-6

Handbook of Hydrogels: Properties, Preparation & Applications
David B. Stein (Editor)
2009. ISBN: 978-1-60741-702-6
2009. ISBN: 978-1-61668-167-8 (E-book)

Treatment of Tannery Effluents by Membrane Separation Technology
Sirshendu De, Chandan Das and Sunando DasGupta
2009. ISBN: 978-1-60741-836-8

Asphaltenes: Characterization, Properties and Applications
Jeremy A. Duncan (Editor)
2010. ISBN: 978-1-60741-453-7
2010.ISBN: 978-1-61668-515-7 (E-book)

Chemicals and Radioactive Materials and Human Development
Rajiv K. Sinha, Rohit Sinha and Shanu Sinha
2010. ISBN: 978-1-61668-145-6
2010. ISBN: 978-1-61668-455-6 (E-book)

Bulk Materials: Research, Technology and Applications

Teodor Frías and Ventura Maestas (Editors)
2010. ISBN: 978-1-60692-963-6

Chromatography: Types, Techniques and Methods
Toma J. Quintin (Editor)
2010. ISBN: 978-1-60876-316-0

Biomedical Chromatography
John T. Elwood
(Editor)
2010. ISBN: 978-1-60741-291-5
2010. ISBN: 978-1-61668-470-9 (E-book)

Flame Retardants: Functions, Properties and Safety
Paulo B. Merlani
2010. ISBN: 978-1-60741-501-5

Bisphenol A and Phthalates: Uses, Health Effects and Environmental Risks
Bradley C. Vaughn
(Editor)
2010. ISBN: 978-1-60741-701-9
2010. ISBN: 978-1-61668-514-0 (E-book)

Green Composites: Properties, Design and Life Cycle Assessment
François Willems and Pieter Moens
(Editors)
2010. ISBN: 978-1-60741-301-1

Photocatalysis on Titania-coated Electrode-less Discharge Lamps
Vladimír Církva and Hana Žabová
2010. ISBN: 978-1-60876-842-4

Environmentally Harmonious Chemistry for the 21st Century
Masakazu Anpo and K. Mizuno (Editors)
2010. ISBN: 978-1-60876-428-0

Perovskites: Structure, Properties and Uses

Maxim Borowski
(Editor)
2010. ISBN: 978-1-61668-525-6
2010. ISBN: 978-1-61668-870-7 (E-book)

Fourier Transform Infrared Spectroscopy: Developments, Techniques and Applications
Oliver J. Rees (Editor)
2010. ISBN: 978-1-61668-835-6

Fluid Phase Behavior of Systems Involving High Molecular Weight Compounds and Supercritical Fluids
Pedro F. Arce and Martín Aznar
2010. ISBN: 978-1-61668-310-8

X-Ray Photoelectron Spectroscopy
Johanna M. Wagner (Editor)
2010. ISBN: 978-1-61668-915-5
2010. ISBN: 978-1-61728-240-9 (E-book)

Preparation of Thin Film Pd Membranes for H2 Separation From Synthesis Gas and Detailed Design of a Permeability Testing Unit
M. Bientinesi and L. Petarca
2010. ISBN: 978-1-60876-538-6

Handbook of Research on Chemoinformatics and Chemical Engineering
A. K. Haghi (Editor)
2010. ISBN: 978-1-61668-504-1

Recent Progress in Chemistry and Chemical Engineering Research
A.K. Hagh (Editor)
2010. ISBN: 978-1-61668-501-0
2010. ISBN: 978-1-61728-237-9 (E-book)

Chemistry and Chemical Engineering Research Progress
A. K. Haghi (Editor)
2010. ISBN: 978-1-61668-503-4

Gadolinium: Compounds, Production and Applications
Caden C. Thompson
2010. ISBN: 978-1-61668-991-9
2010. ISBN: 978-1-61728-334-5 (E-book)

CHEMICAL ENGINEERING METHODS AND TECHNOLOGY

PREPARATION OF THIN FILM PD MEMBRANES FOR H2 SEPARATION FROM SYNTHESIS GAS AND DETAILED DESIGN OF A PERMEABILITY TESTING UNIT

M. BIENTINESI
AND
L. PETARCA

Nova Science Publishers, Inc.
New York

For permission to use material from this book please contact us:
Telephone 631-231-7269; Fax 631-231-8175
Web Site: http://www.novapublishers.com

LIBRARY OF CONGRESS CATALOGING-IN-PUBLICATION DATA

Bientinesi, M. (Miguel), 1950-
Preparation of thin film Pd membranes for H2 separation from synthesis gas and detailed design of a permeability testing unit / M. Bientinesi, L. Petarca.
p. cm.
Includes bibliographical references and index.
ISBN 978-1-60876-538-6 (hardcover)
1. Hydrogen--Separation. 2. Palladium. 3. Membranes (Technology) 4. Gasoline, Synthetic. I. Petarca, L. (Luca), 1945- II. Title.
TP359.H8B535 2009
665.8'1--dc22

2009041957

Published by Nova Science Publishers, Inc. ✦ New York

CONTENTS

Preface		**xi**
Chapter 1	Introduction	**1**
	1.1. Hydrogen as an Energy Carrier	*1*
	1.2. Current Production Methods and Uses	*2*
	1.3. Syngas for Hydrogen Production	*5*
	1.4. Separation of Hydrogen from Synthesis Gas	*8*
	1.5. Fundamentals of Hydrogen Membrane Separation Technology	*11*
Chapter 2	Palladium Membranes	**25**
	2.1. Pure Palladium and Palladium Alloys Membranes	*26*
	2.2. Porous Supports	*28*
	2.3. Palladium Film Deposition Techniques	*31*
	2.4. Palladium Membranes Characterization	*37*
Chapter 3	Experimental	**39**
	3.1. Metallic Porous Supports	*39*
	3.2. Supports Abrasion and Cleaning	*39*
	3.3. Surface Oxidation	*40*
	3.4. Sensitization and Activation	*40*
	3.5. Electroless Plating	*41*
	3.6. Morphology Characterization	*46*
	3.7. Nitrogen Low Pressure Permeation Test	*46*
Chapter 4	Result and Discussion	**49**
	4.1. Porous Supports	*49*
	4.2. Support Abrasion And Etching	*50*

	4.2. Surface Oxidation	*51*
	4.3. Palladium Deposition	*52*
	4.4. Tubular Membranes Nitrogen Permeability	*57*
Chapter 5	Permeability Test Experimental Set-Up	**61**
	5.1. Introduction	*61*
	5.2. Experimental Set-Up	*62*
	5.3. Selected Equipment	*64*
	5.4. Permeation Cells and Other Equipment	*65*
Chapter 6	Conclusion	**67**
Acknowledgements		**69**
Bibliography		**71**
Index		**75**

PREFACE

Hydrogen is one of the most important chemical products, and has uses in a wide series of industrial fields: chemical, petrochemical, metallurgical, as well as in energy applications. In all these fields, the demand for hydrogen is continuously growing. Nowadays hydrogen is mainly produced from fossil fuels via such processes as steam reforming, partial oxidation, and gasification. Many of these processes lead to the production of a gas stream (synthesis gas) from where hydrogen is then separated.

Thin-film palladium membranes are one of the most promising technologies for the separation of hydrogen from synthesis gas. It involves some advantages over traditional separation methods like pressure swing adsorption (PSA), and other membrane materials (polymeric, porous or dense ceramic). Advantages include a theoretic 100% separation efficiency, high permeability, and operating conditions compatible with upstream fuel conversion processes.

In this work, palladium films were deposited above stainless steel porous supports using the electroless plating (ELP) technique. The supports have a 0.1 μm filter grade. Two different geometr ies (disc sheet and tubes) were used.

Surface morphology and cross-section were observed through scanning electron microscopy (SEM). For each membrane, film thickness was estimated both by weight gain and by cross-section observation with SEM. A good comparison was found between the two values. The evolution of film thickness and morphology was studied for the increasing number of ELP cycles, as well as the influence of two different pre-treatments (surface abrasion and thermal oxidation) of the metallic substrate.

Membranes were tested in an appropriate set up for nitrogen tightness, in order to isolate the possible presence and the localization of defects. The results showed that after only 4 cycles of deposition, a uniform dense film of palladium with a thickness of about 10 μm is obtained, but even after 6 cycles, a small number of defects still occur. These defects were found to be due to local morphological discontinuities in the support surface. As a consequence, the pre-treatment of abrasion lead to a smoother support surface that allowed for a t reduction in the number of defects, and consequently a reduction in nitrogen flux for the same number of deposition cycles.

Finally, a detailed design of the permeability testing unit is reported that includes the fluid dynamic, thermal and mechanical dimensioning, the selection of materials and equipment, and some safety considerations. The sketches of two permeation cells (for two different membrane geometries) are also described.

Chapter 1

INTRODUCTION

This introduction provides an overview for the current status in the use of hydrogen as an energy carrier, and includes a description of its production methods and utilization technologies. Attention is focused on theprocesses of fossil or renewable fuels thermal conversion that produce synthesis gas from where hydrogen is then separated.

Different hydrogen separation technologies from gas mixture are reviewed, with particular concern to membrane technologies.

1.1. HYDROGEN AS AN ENERGY CARRIER

Hydrogen is often erroneously considered as a primary source of energy. This is absolutely wrong, because we cannot extract hydrogen from nature, but can only produce it from other available resources. Hydrogen has to therefore be considered as an energy carrier; or, in other words, viewed as an energy transfer medium, in the same way as electricity (Baade et al., 2001, Collot, 2003).

The use of hydrogen as energy carrier allows a series of important advantages with respect to, for instance, electricity:

- hydrogen can be stored and transported with no significant losses;
- H_2 combustion can be managed in order to be a zero-emission process (for example in fuel-cell) and it does not involve carbon dioxide emissions;

- it can be used in large power plants as well as in small size, fixed or mobile power or CHP (combined heat and power) generation units.

The concept of the "Hydrogen Economy" was introduced in the '70s (Gregory & Pangborn, 1976): in this kind of scenario hydrogen has to be produced from renewable resources, in order to completely eliminate carbon dioxide emissions. However renewable technologies are not yet sufficiently widespread to fulfil this goal.

In the near future, it is therefore more realistic to employ fossil resources to produce hydrogen via well-established technologies such as methane steam-reforming or coal gasification. This approach will clearly involve some important advantages with respect to the direct utilization of fossil fuels in power plants or mobile applications:

- the possibility of capturing and storing carbon dioxide in order to limit the greenhouse effect;
- the possibility of using hydrogen in innovative highly-efficient applications such as fuel-cells;
- for automotive applications, the substitution of petrol and diesel with hydrogen (burned in fuel-cell engine or even directly in internal combustion engines); this is particularly important because it allows the delocalization of polluting emissions from urban centers to H_2 production plants, this provides more efficient abatement systems;
- the fact that hydrogen can be burned in catalytic combustor, thanks to its low ignition energy; catalytic combustion takes place at low temperature (500°C), thus it is safe and NO_x formation is almost completely eliminated; moreover it can be used even in small size domestic applications (Stephan & Dahm, 1996);
- the immediate availability of hydrogen at low prices, could generate a propulsive push to the development of infrastructures and equipment which are indispensable for a gradual move towards the use of H_2 as energy carrier.

1.2. Current Production Methods and Uses

Current worldwide production of hydrogen is around $4.5 \cdot 10^{10}$ kg/yr (US DOE, 2002). Most of this production is immediately consumed on site (for example in refineries), while only a minor part (about 10%) is produced to be transported elsewhere and sold to other companies (Meteyer, 2008).

The main sources of hydrogen are gaseous, liquid or solid fossil fuels, from which H_2 production takes place via well established processes of thermal conversion (steam reforming, gasification, and partial oxidation). On the other hand, water electrolysis and biomass gasification are still relatively marginal technologies (**Figure 1**). Several other H_2 production methods are under development, but have not yet found industrial applications. Finally, hydrogen is also an important co-product of many other chemicals (ethylene, styrene, MTBE production) and petrochemical (petrol catalytic reforming) processes.

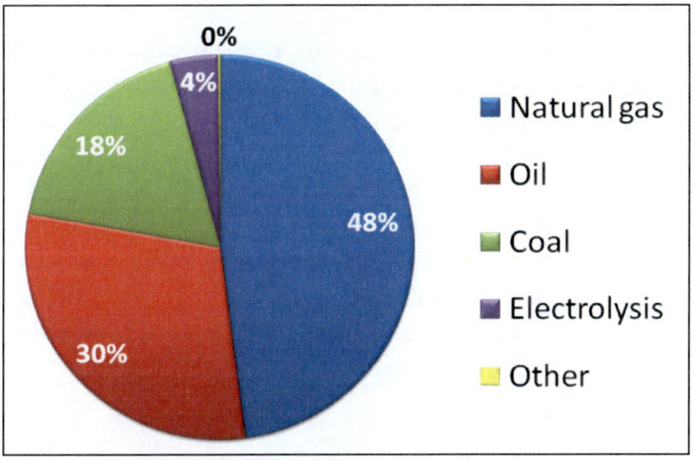

Figure 1. Main sources for hydrogen worldwide production (2002).

Up to now, hydrogen utilization as energy carrier has been limited by the difficulty of transporting this gas due to the low energy density, to its explosiveness, flammability, and volatility. Hydrogen can be stored and transported in several ways (Baade et al., 2001). Besides the traditional technologies such as compressed gas and cryogenic liquefied hydrogen, newer technologies, such as metal hydrides, glass microspheres and carbon nanotubes or nanofibers storage, are under development. They tend to improve storage and transport safety while maintaining an high energy density and low power consumption.

For the capillary distribution of hydrogen towards buildings, hydrogen pipelines or modified natural gas pipelines can be used. In Europe it was estimated that about 1100-1800 km of hydrogen pipelines have been built and are currently used; mainly by the chemical industry, while in the USA the existing H_2 transmission system ranges from 700 to 1300 km. As a way of comparison, in the USA there exist about 300000 km of natural gas pipelines (Gillette & Kolpa, 2007).

Some of hydrogen's peculiarities contribute to the increase of its distribution costs. First of all, hydrogen, when put in contact with steel, promotes embrittlement that can lead to pipes and gaskets rupture; so special materials and solutions are required. Moreover, explosion-proof systems have to be used.

Hydrogen cost is currently dramatically higher than the cost of other fuels such as petrol, diesel and natural gas; primarly because it is produced only on a very small scale. Its use as a fuel is now limited to the propulsion of the space shuttle and to some circumscribed experimentations.

The commercial production of pure hydrogen started at the beginning of the 20th century with the development of ammonia and methanol synthesis (Baade et al., 2001). These two processes together consume about three-quarters of the hydrogen worldwide production, while the remaining is used in oil product refining, metallurgy, as well as in the electronic and food industry (**Figure 2**). The portion used by refineries for the processes of hydrotreating and sulphur removal from fuels, currently represent almost a quarter of H_2 production. This portion is expected to grow sharply in the next few years, due to the increase of oil derived products consumption, to the lower quality of the extracted oil, and to the increasingly strict legislation of many countries about sulphur, nitrogen and aromatics content in fuels.

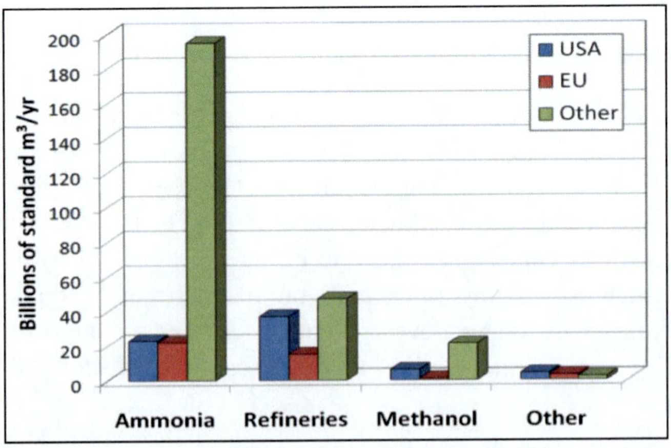

Figure 2. Hydrogen use by end-market and by country (2001).

Hydrogen as a fuel can be used in different applications, that can be classified in:

- fixed applications (power generation, cogeneration, trigeneration);
- mobile applications (automotive, railway, naval and airplane transport);
- portable applications (mobile phones, laptops, other electronic equipment batteries).

On the basis of the technology used, we can distinguish:

- fuel cells are the most promising technologies with potential applications in all the described fields;
- hydrogen fuelled internal combustion engines (automotive traction);
- Stirling engines and microturbines (small cogenerative units);
- hydrogen/methane mixtures use in the municipal gas distribution network (domestic and commercial heating).

Many of the described technologies are still under development, or have costs still too high for commercial distribution. Anyway, several car companies have for instance developed a hydrogen fueled internal combustion car (BMW, Mazda) or a fuel cell car (Audi, Chrysler, Daimler, Fiat, Ford, GM, Honda, Hyundai, Mazda, Nissan, Peugeot, Renault, Toyota, Volkswagen). In some cases these cars are already in commerce.

1.3. Syngas for Hydrogen Production

In **Figure 1** we reported that about 96% of the worldwide hydrogen production comes from fossil fuels; namely coal, oil, and natural gas. In all these cases, the first step is the conversion of the primary resource to a gas mixture. This gas mixture is called synthesis gas or syngas, and is primarily composed of H_2, CO, CO_2, CH_4, N_2 and other minor components, with variable composition (Shaole et al, 2001, Haussinger et al., 2003, Collot, 2003). A large number of different processes, dating back to 18th century, have been used for this purpose.

Steam reforming of natural gas (or **steam methane reforming**) is an endothermic process in which methane reacts with steam to give carbon monoxide and hydrogen, according to the reaction (performed in catalytic beds):

$$CH_4 + H_2O \leftrightarrow CO + 3H_2 \quad \Delta H^0_{298K} = 206.16\ kJ/mol \tag{1}$$

The heat of reaction is usually furnished by the combustion of natural gas. Usually, when hydrogen is the final product or a high hydrogen content syngas is required, this reactions is followed by another one, called *water gas shift reaction*, which can take place in apposite catalytic reactors:

$$CO + H_2O \leftrightarrow CO_2 + H_2 \quad \Delta H^0_{298K} = -41.16\ kJ/mol \tag{2}$$

The overall reaction turns out to be:

$$CH_4 + 2H_2O \leftrightarrow CO_2 + 4H_2 \tag{3}$$

The overall cold gas efficiency defined as the ratio of energy content, on HHV basis, of the produced gas to the energy content of the primary fuel used, ranges from 70% to 85%. In those regions where natural gas is not easily available, naphtha or other light hydrocarbons with boiling temperature between 44°C and 56°C are used to feed the steam reforming process. In this case, lower efficiency and lower H_2/CO ratio are obtained.

Another way to convert natural gas or liquid hydrocarbon into syngas is by partial oxidation, in which substoichiometric oxygen or air is fed in order to perform the general reaction:

$$C_nH_m + \frac{n}{2}O_2 \rightarrow nCO + \frac{m}{2}H_2 \tag{4}$$

Compared to steam reforming, cold gas efficiency is largely lower (about 60%) but no catalyst is needed. Even in this case, water gas shift reaction can be performed in a second reactor to maximize the hydrogen yield.

In coal gasification, coal reacts with gaseous species such as O_2, H_2O, CO_2 to form a product gas rich in carbon monoxide and hydrogen. It consists essentially of a partial combustion that can be conducted in several different conditions; for example, feeding only sub-stoichiometric air, or oxygen, or adding also steam; moreover it can be performed in direct or indirect gasifiers, and in fixed bed, fluidized bed or entrained flow gasifiers; finally different operating conditions (reagents ratios, temperature, pressure) can be applied. Different process options, as well as different coals, lead to different coal conversions (90-99%), to different efficiencies and to different syngas compositions, so they have to be selected on the basis of the syngas final utilization. Oxygen and steam are used as reagents when hydrogen is the desired final product. In any case, the main chemical reactions involved in the process are the following:

$$(1+\lambda)C + O_2 \rightarrow 2\lambda CO + (1-\lambda)CO_2$$
$$\Delta H^0_{298K} = (172.5\lambda - 393.5)\ kJ/mol\ (0<\lambda<1) \quad (5)$$

$$C + CO_2 \leftrightarrow 2CO\ \ \Delta H^0_{298K} = 172.5\ kJ/mol \quad (6)$$

$$C + H_2O \leftrightarrow CO + H_2\ \ \Delta H^0_{298K} = 131.3\ kJ/mol \quad (7)$$

$$C + 2H_2 \leftrightarrow CH_4\ \ \Delta H^0_{298K} = -74.8\ kJ/mol \quad (8)$$

Sulphur contained in the coal is converted mainly to H_2S and COS, which can be quite easily removed by the syngas, while nitrogen is converted to N_2 plus little amount of NH_3. No SO_x and NO_x form. Again, the process can be followed by the water gas shift reaction, and overall cold gas efficiency ranges from 50% to 65%.

In addition to coal gasification, **biomass gasification** can be a suitable process for the production of hydrogen. It has the additional advantage of being CO_2 neutral (because the CO_2 deriving from organic carbon cannot be considered as a greenhouse gas emission), and giving a higher hydrogen yield thanks to the higher H content of biomass with respect to coal. On the other hand, thermal efficiency is low (35-50%), due to the high moisture content of the fuel. Problems may arise from the high tar and dust content of biomass gasification derived syngas, which may be difficult to remove.

Inall of the described processes, after the shift reaction and the necessary gas cleaning operations, a gas stream extremely rich in hydrogen (50-75% by volume) and carbon dioxide is obtained. For instance, in **Table 1** typical range compositions of a natural gas steam reforming and a coal gasification derived syngas, after water gas shift reaction and gas cleaning, are reported. At this point, to obtain a pure hydrogen stream, CO_2 and other residual gases have to be separated and hydrogen has to be compressed in order to be sent to the storage or directly to the utilization (**Figure 3**).

Table 1. Hypothetical average composition of clean syngas.

	NG steam reforming	Coal gasification
H_2 [vol %]	75	55
CO_2 [vol %]	15	35
CO [vol %]	5	5
H_2O [vol %]	1	4
CH_4 [vol %]	4	1
Minor components	N_2, H_2S, COS, NH_3	

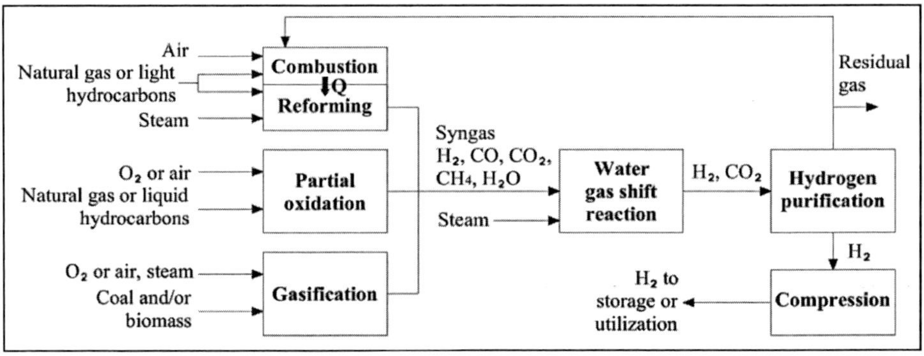

Figure 3. Hydrogen production processes from fossil fuels and biomass.

1.4. Separation of Hydrogen from Synthesis Gas

Hydrogen can be separated from other components of syngas mainly through two different technologies:

- pressure swing adsorption (PSA);
- membrane technologies.

1.4.1. Pressure Swing Adsorption

Pressure swing adsorption has been used in refineries and chemical plant since the ‘60s of past century (Baade et al., 2001). This process is based on the capacity of solid sorbent such as zeolite or activated carbon to adsorb, with a certain selectivity, a high amount of determined compounds (adsorbates) at high pressure, and to release them at low pressure (Yon & Sherman, 2003). In particular, hydrogen is much poorly adsorbed compared to heavier syngas components, and particularly to CO_2. In a hydrogen purification PSA cycle, therefore, CO_2 (and other minor components) adsorption is performed at high pressure, while desorption is performed at low pressure. The bed regeneration phase requires two steps: depressurization and purging. Depressurization lowers the CO_2 partial pressure in order to thermodynamically consent its desorption, while purging, performed feeding a stream called tail gas, transports the desorbed gas away from the vessel. The cycle is adiabatic, because the heat generated during adsorption rests in the sorbent and is consumed during desorption. The process is carried out in a continuous way by using 4 or more beds, each performing one step (adsorption, depressurization, purging, re-pressurization) of the cycle.

The driving force for the separation is the difference in adsorbates partial pressure between feed and tail gas. In order to obtain a higher purity of hydrogen, the ratio has to be of 4 to 1 at least. Absolute pressure of the feed stream is usually in the range 15-30 bar, while tail gas must be as low as possible (vacuum is usually avoided, and a nearly atmospheric pressure is used). As tail gas, a small part of the produced hydrogen is employed. The adsorbates containing stream is then burned to recover its energy content.

Pressure swing adsorption is a well established technology that can accomplish the production of hydrogen with a higher purity (from 99% to 99.999%). Moreover, the level of impurities can be tuned by simply changing operating conditions, without significantly impacting on hydrogen recovery. On the other hand, hydrogen recovery is limited, and varies from 80% to 92% in optimal conditions.

1.4.2. Membrane Separation

Hydrogen separation membranes have been used in refineries and petrochemical plant since the end of ‘70s. In the last 20 years, the interest in this field has grown significantly, as can be seen in **Figure 4**, where the number of publications on scientific journal and of patents on this topic is reported. This interest is determined by a series of advantages that membrane separation of hydrogen guarantees respect to PSA (Grimmer et al., 2006):

- higher hydrogen recovery;
- hydrogen purity up to 100% for some types of membrane;
- lower operating cost;
- it does not involve mobile parts susceptible of rupture;
- the separation plant capacity can be easily expanded by adding new membrane modules.

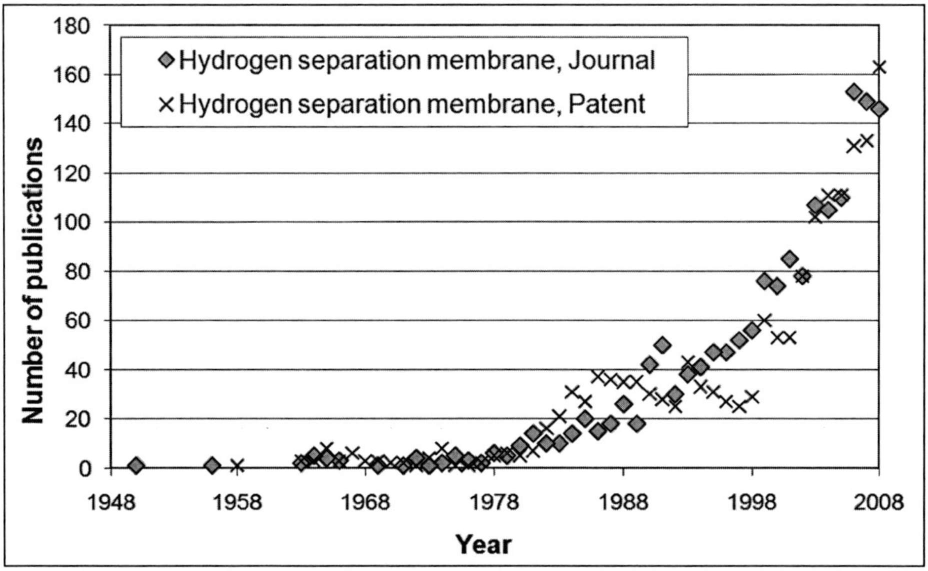

Figure 4. Number of publications per year about hydrogen separation membranes.

Another important advantage can derive from the fact that in the near future, in order to limit greenhouse gas emissions; it will be more and more necessary to sequestrate CO_2 streams generated by fossil fuels conversion processes. If PSA is used to separate H_2 from syngas, produced CO_2 is not pure and is at a pressure slightly above atmospheric pressure. For almost all membrane process separation, on the other hand, CO_2 is produced practically at the feeding pressure, which can be up to 35 bar. The CO_2 compression from 35 bar to 70 bar (sequestration pressure) requires only 1/6 of the energy required by the compression from 1 bar to 70 bar, leading to important economic and environmental benefits.Some fundamental properties of the two separation processes are compared, see Table 2.

Table 2. Comparison of H_2 purification via membrane separation and PSA.

	MEMBRANES	PSA
H_2 purity [%]	95 – 100	99.9
H_2 recovery [%]	>90	75-90
Feeding pressure [bar]	14 – 50	10 – 100
H_2 product pressure	Atmospheric pressure (or feeding pressure)	Feeding pressure

1.5. FUNDAMENTALS OF HYDROGEN MEMBRANE SEPARATION TECHNOLOGY

Membranes for gas mixture separation are made of materials characterized by a permeation rate specific for every single gas compound (Koros et al., 1996). This characteristic allows the separation between a first stream, called **permeate**, which is rich in those gases that cross more rapidly through membrane material; and a second stream, called **retentate**, that contains residual gases.

The most important characteristics of gas separation membranes are:

- **permeability** (in $mol \cdot m/m^2 \cdot s \cdot Pa$ or similar units), is the capacity of a specific compound to permeate the membrane material and is calculated as the gas molar flux through the membrane, normalized with respect to membrane area, membrane thickness and transmembranal gas partial pressure gradient;
- **permeance** ($mol/m^2 \cdot s \cdot Pa$), which is a characteristic of the specific membrane and not of the membrane materials (such as permeability); in this case the flux is normalized with respect to membrane area and transmembranal gas partial pressure gradient only; permeance is often used in place of permeability when membrane thickness is uncertain or unknown;
- **selectivity**, defined as the ratio between the permeabilities (or permeances) of two different gaseous species that have to be separated; in practice it represents the membrane separation efficiency;
- **operating conditions**,can be expressed as the range of temperature and feeding and permeate pressure in which the membrane can be correctly operated with acceptable performances and without breaking or deterioration risks;

- **chemical resistance** towards molecules contained in the feeding gas, which can sometimes deteriorate membrane performances or even cause dramatic structural changes in the material.

It is worthy to note that membrane performances vary strongly with temperature and other operating conditions. Therefore the selection between the different membrane types should be done taking into account the process where the membrane module has to be inserted and the final destination of the produced hydrogen, which strongly influences product purity requirements.

A first classification between membrane modules can be made on the base of geometry, which can vary depending on membrane material and application. Four geometries are possible:

- tubular;
- flat sheet or plate frame;
- hollow fiber;
- spiral wound.

In any case, the driving force is a difference in the partial pressure of each gas between the feeding side (high partial pressure) and the permeate side (low partial pressure). This difference can be inducted by feeding an inert sweep gas on the permeate side or, more frequently, by applying an absolute pressure gradient between the two sides of the membrane. Many different membrane materials, with different transport mechanisms, have been developed and tested for hydrogen separation. These can be classified in (Pandey & Chauhan, 2001):

- dense polymeric membranes;
- dense metallic membranes;
- ceramic microporous membranes;
- dense ceramic membranes;
- composite membranes.

1.5.1. Dense Polymeric Membranes

The transport mechanism of gases through dense polymeric membrane is called solution-diffusion mechanism (Robeson, 1991, Matteucci et al. 2006). It involves three phases:

- molecule adsorption on feed side;
- activated diffusion of the molecule through the polymer under a partial pressure gradient;
- molecule desorption on permeate side.

The passage of molecules is due to the intrinsic microporosity of the polymer structure. The flux of a molecule A through a unit area of membrane can be calculated as follows:

$$J_A = \frac{P_A\left(p_A^{feed} - p_A^{perm}\right)}{s} \tag{9}$$

Where, p_A^{feed} e p_A^{perm} [Pa] are the partial pressure of A respectively in the feed side and in the permeate side, s [m] is the membrane thickness, P_A [mol·m/m^{2}sPa] is the permeability of A through the polymer. The permeability for a dense polymeric material can be expressed as the product of a diffusivity coefficient D [m^2/s] and of a solubility coefficient S [mol/m^{3}Pa]:

$$P_A = D_A \cdot S_A \tag{10}$$

Thus the selectivity of molecule A with respect to molecule B is:

$$\alpha_{A/B} = \frac{P_A}{P_B} = \frac{D_A}{D_B} \cdot \frac{S_A}{S_B} \tag{11}$$

The diffusivity coefficient depends essentially on the molecule dimension, and it decreases with increasing molecule dimension. For several polymers, it can be fitted versus the critical volume of the gas according to an inversely proportional relation. The solubility coefficient, on the other hand, is higher for easily condensable gases, and can be related to their critical temperature (Matteucci et al. 2006, Freeman & Pinnau, 1997).

The main part of the commercially available polymeric membranes exploit diffusion selectivity, thus permeating small molecules, while blocking bigger ones. They are made of materials such as polyimide, polysulfone, and cellulose acetate. These membranes are capable of separate H_2 from a series of compounds such as CO, CH_4, N_2, H_2S, C_2-C_5, Ar with good selectivity, and are therefore largely employed for hydrogen recovery from purge stream in petrochemical, ammonia or methanol production plants, as well as for the adjustment of H_2/CO ratio of syngas. Unfortunately, because of the high solubility coefficient of CO_2, they have a really low H_2/CO_2 selectivity ($D_{CO2}/D_{H2}>>1$, $S_{CO2}/S_{H2}<<1$), ranging from 2 to 3, and cannot be used for the separation of hydrogen from carbon dioxide (Baker, 2002).

In order to overcome this limit, other polymeric material are being studied that have a very high CO_2 solubility and a low molecular sieve effect, so bringing the ratio D_{CO2}/D_{H2} as near to 1 as possible. In this way, CO_2 turns out to be the more permeable gas and goes into the permeate, while most of H_2 remains in the retentate, with a selectivity up to 10 and even more (Lin et al., 2006). This system could be promising especially for applications that require downstream high pressure hydrogen and do not perform carbon dioxide sequestration, but they are still in an early stage of development.

Polymeric membrane modules are essentially produced in two geometries, hollow fiber and spiral wound, that both aim for a high surface area in a small volume. Usually membrane modules are composed of a thin dense skin layer, with thickness below 0.5 μm, deposited on a thicker porous layer of the same material. The skin layer act as a selective membrane, while the porous layer provides mechanical strength. Other configurations can use up to 3 or 4 layers of different, less expensive materials (Baker, 2002).

Polymeric membranes are operated at temperature lower than 100-150°C, and can support transmembrane pressure up to 100 bar. Hydrogen permeability is relatively low, in the order of 10^{-5} $mol \cdot m/m^2 \cdot s \cdot bar$, leading to the necessity of high surface area configurations.

The interest towards polymeric membrane has recently declined, due mainly to problems of fouling and thermal stability, that strongly limit the possibility of their application toward separation processes involving hot and reactive gases (Pandey & Chauhan, 2001).

1.5.2. Dense Metallic Membranes

Selective dissolution of hydrogen in metallic palladium has been around for a very long time (Graham, 1866), as well as in other metals such as Ag, Ti, Nb, V, Ta, Fe, Ni, Pt (Wipf, 2001, Adhikari & Fernando, 2006). Research has focused on Pd because this metal has the best performances in terms of permeability and selectivity coupled with thermal and chemical resistance. Hydrogen permeation through dense palladium layers implies a number of steps in series, according to the "solution-diffusion" mechanism depicted in **Figure 5** (Pandey & Chauhan, 2001, Sholl & Ma, 2006, Ward & Dao, 1999):

a) the molecular transport from the bulk gas to the metal surface on the feed side;
b) the dissociative adsorption of the H_2 molecule on the metal surface (1);
c) the dissolution of the H atom into the metal lattice (2);
d) the atomic diffusion into the bulk metal (3);
e) the transition from the bulk metal to the metal surface on the permeate side (4);
f) the recombinative desorption from the surface;
g) the molecular transport away from the metal surface towards the bulk gas.

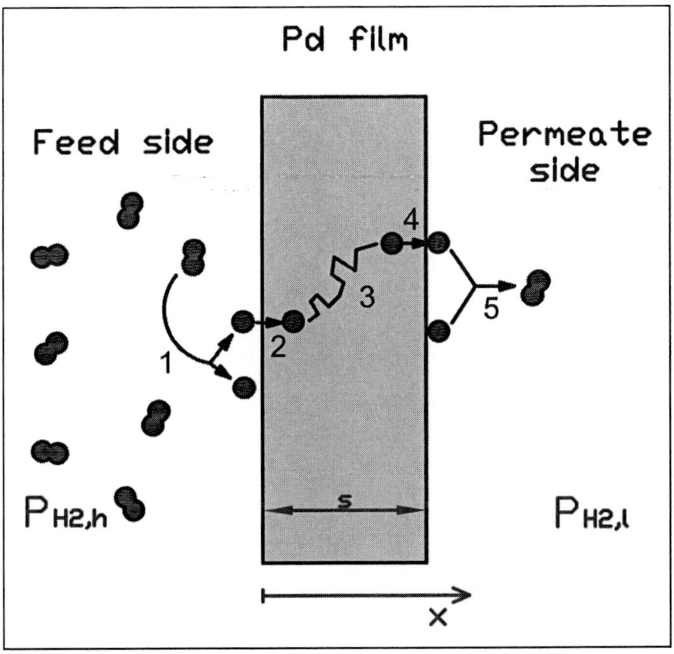

Figure 5. H_2 permeation through metallic Pd.

Each step has a proper rate, and the overall H_2 transport rate can be limited by one or more of these steps. Molar hydrogen flux through the metal can be expressed by the Fick's first law:

$$J_{H2} = D_{H2}\frac{dC_H}{dx} \tag{12}$$

Where, D_{H2} is the diffusion coefficient, C_H is the hydrogen atomic concentration in the metal lattice and x is the spatial coordinate in the H_2 flow direction. If we assume that adsorption, dissociation, and dissolution are very rapid (for instance at high temperatures), H_2 flow is limited by atomic diffusion in the bulk metal (step 3) and Sievert's law is followed (Athayde et al. 1994):

$$C_H = K \cdot p_{H2}^{\ 0.5} \tag{13}$$

This law expresses the equilibrium at the gas solid interface between the H_2 partial pressure in the gas phase (p_{H2}) and H molar concentration in the metal (C_H). By assuming D_{H2} and K constant and by integrating on the full membrane thickness, we have:

$$J_{H2} = \frac{KD_{H2}}{2}\frac{\sqrt{p_{H2,h}} - \sqrt{p_{H2,l}}}{s} = \frac{P_{H2}}{s}\left(\sqrt{p_{H2,h}} - \sqrt{p_{H2,l}}\right) \tag{14}$$

where P_{H2} [mol·m/m^2·hr·bar$^{0.5}$] is the hydrogen permeability, s [m] is the membrane thickness, $p_{H2,h}$ and $p_{H2,l}$ [bar] are the hydrogen partial pressure respectively on the feed and on the permeate side. A more general expression is usually adopted (Sholl & Ma, 2006, Ward & Dao, 1999, Tong et al., 2004):

$$J_{H2} = \frac{P_{H2}}{s}\left(p^n_{H2,h} - p^n_{H2,l}\right) \tag{15}$$

where n is an experimentally determined exponent. If the atomic diffusion in the bulk metal (step 3) is the limiting step, n is found out to be near 0.5, in agreement with Sievert's law. On the contrary, n assumes growing values between 0.5 and 1 if surface phenomena (adsorption, dissolution) become slower and kinetically relevant.

The flow dependence from temperature can be modeled with the Arrhenius' law:

$$P_{H2}(T) = P_{H2,0} \cdot \exp\left(-\frac{E_{att}}{RT}\right) \tag{16}$$

where E_{att} (J/mol) is an apparent activation energy.

The advantages of palladium membranes are theoretically infinite selectivity and high permeability. On the other hand, they still have some problems. At temperature lower than 300°C and in the presence of hydrogen, palladium exhibits a phase transition α-β. The consequent lattice expansion and contraction can lead to the cracking of the membrane when it undergoes temperature cycles. This problem can be avoided in two ways: avoiding the presence of hydrogen when the membrane is cooled (for instance purging with nitrogen before stopping separation) or using palladium alloys with metal such as copper and silver, which do not undergo phase transition.

Palladium based membranes, with tubular geometry, are currently commercialized by two different producers (Johnson Matthey, Pd/Ag alloy, and ATI Wah Chang, Pd/Cu alloy). They are relatively thick (25-100 μm) membranes with no support, made through the rolling method. Due to the high thickness, they are expensive and have a limited mechanical strength. They are used mainly for the production of high purity hydrogen (impurities at ppb levels) for special applications (fuel cells, semiconductor industry). They operate at temperature between 300°C and 600°C, with transmembranal pressures between 20 bar and 40 bar. The catalytic nature of palladium leads to a high sensitivity towards compounds such as sulphur compounds, that react with surface Pd altering metal morphology (irreversible poisoning); and as unsaturated hydrocarbons, Hg, Cl_2, HCl, which are chemisorbed on the surface (reversible poisoning). Hydrogen flow reachable with these commercial membranes is low, due to their relatively high thickness, and it varies from 0.01 to 0.03 $mol/m^2 \cdot s \cdot bar$. For this reason they cannot be applied for hydrogen production in the energetic field.

Many works aim for the improvement of Pd-based membranes performances. Two main research directions are followed. The deposition of thin dense selective metallic films, with thickness ranging from 0.2 μm to 20 μm, above porous supports that assure mechanical resistance, leads to two important advantages: flow increase and cost lowering. Supports can be made of different materials, both ceramic and metallic. The second important research line is the use of palladium alloys, that in many cases allows to obtain, in addition to thermal stability, higher permeability, lower poison sensitivity, and higher mechanical strength.

Dense palladium membranes production techniques are discussed in section 2.

1.5.3. Microporous Ceramic Membranes

Five different mechanisms are involved in gas transport in porous membranes (Pandey & Chauhan, 2001, Uhlhorn et al., 1992, De Lange et al., 1995):

- viscous or Poiseuille flow;
- Knudsen diffusion;
- capillary condensation;
- surface diffusion;
- molecular sieving.

The relative contribution of each mechanism to the gas permeation depends on membrane characteristics, as well as on the gas mixture, and the operating conditions (i.e., temperature, pressure).

Viscous flow is non- selective, and is predominant when the molecule mean free path is largely lower than the pore radius. **Knudsen diffusion**, on the contrary, predominates for pore radius smaller than the molecule mean free path. This condition is encountered in membranes with pore diameter smaller than 50 nm. Knudsen flow involves a series of collision between gas molecules and pore walls, during which molecules are adsorbed and reflected in casual directions. The flow of different gases is independent, and the separation of gas mixtures can be obtained because different molecules move at different rates. The permeation rate is approximately inversely proportional to the square root of themolecular weight. Thus the ideal selectivity between two molecules A and B can be calculated by:

$$\alpha_{A/B} = \sqrt{\frac{M_B}{M_A}} \tag{17}$$

Where, M_A and M_B are the molecular weights of two gases. Ideal Knudsen selectivity for some hydrogen-containing mixtures are reported in **Table 3**. In fact, actual separation factors are lower than the reported ideal values, because of mechanisms such as back-diffusion, non-separative diffusion, concentration polarization, and viscous flux in larger pores.

Table 3. Calculated Knudsen selectivity for H_2-containing mixtures.

Mixture	Separation factor
H_2/N_2	3.74
H_2/CO	3.74
H_2/H_2S	4.33
H_2/CO_2	4.69

Membranes with pore diameter between the diameters of the two molecules to be separated act as **molecular sieves**, and are characterized by theoretical infinite selectivity. In fact, such membranes are difficult to produce due to the extremely low required pore dimension (see **Table 4** for gas kinetic diameters), and selectivities higher than 10 can be obtained with pores of 0.5 nm.

Table 4. Kinetic diameter of gas molecules (Breck, 1973).

Gas	Kinetic diameter [nm]
H_2O	0.265
H_2	0.289
CO_2	0.33
N_2	0.364
CH_4	0.38

Gas separation can be also influenced by partial condensation of some species (**capillary condensation**) and the subsequent transport of the condensed molecules in the pores. Another important phenomenon can be the selective adsorption of some components of the mixture on the pore surface, followed by **surface diffusion**. These two mechanisms are usually counterproductive for hydrogen separation, since H_2 is a hardly condensable and poorly adsorbed gas compared with, for instance, CO_2.

Porous membranes are actually characterized by a distribution of pore dimensions; thus the gas permeability is influenced by the combination of different transport mechanism. Lowering the mean membrane pore dimension, porosity is lowered, and a lower gas flux is obtained. Thus in order to obtain an efficient membrane, porosity and pore dimension or, in other words, permeability and selectivity, have to be accurately balanced.

According to IUPAC (Sing et al., 1985), porous membranes can be classified on the basis of pore dimension in:

- macroporous membranes ((d > 50 nm);
- mesoporous membranes (2 nm < d < 50 nm);
- microporous membranes (d < 2 nm).

For commercial membranes, with pore dimension above 4 nm, Knudsen diffusion is the main transport mechanism, and low selectivities are obtained for the separation of hydrogen from low molecular weight gases. In recent publications (Prabhu & Oyama, 2000, Asaeda & Yamasaki, 2001, Nomura et al., 2005) it has been reported the production of microporous membranes with selectivities H_2/N_2 as high as 1000 or more.. Experimentally, it has been found that the flux of a gas A (J_A) in these membranes is proportional to the difference of A partial pressure between the two sides of the membrane:

$$J_A = \frac{P_A}{s}\left(p_{A,h} - p_{A,l}\right) \tag{18}$$

and that the permeability P_A ($mol \cdot m/m^2 \cdot s \cdot Pa$) grows with the temperature according to Arrhenius law (De Lange et al., 1995):

$$P = P_0 \exp\left(-\frac{E_{att}}{RT}\right) \tag{19}$$

with apparent activation energies in the range 2-40 kJ/mol, depending on membrane pore dimension and on the permeating gas molecule. The transport mechanism cannot be viscous flux, which is proportional to 1/T; nor Knudsen diffusion, proportional to $1/\sqrt{T}$; nor surface diffusion, which decreases with increasing temperature. So the transport mechanism in microporous membranes has been called activated transport or **microporous diffusion** (De Lange et al., 1995). The selectivity is determined by a series of factors, such as dimension, adsorption characteristics, and microporous diffusivity of molecules.

Ceramic microporous membranes for hydrogen separation are composed by a certain number of layers of different materials. Usually one or more very thin layers with pore dimension below 1 nm are deposited above a macroporous ceramic or metallic support, which supplies mechanical strength. The support can be composed by different layers with variable pore dimension, too, in order to facilitate the above film adhesion. Microporous films are generally made of metallic oxide, namely silica (SiO_2), γ-alumina (γ-Al_2O_3), titania (TiO_2). The supports are usually made of γ or α-alumina, vycor glass, or even stainless steel. Ceramic films are deposited mainly by two techniques, *chemical vapor deposition* or *sol-gel technique*, and their thickness varies from 0.01 to 10 μm.

The use of this type of membrane for hydrogen purification present some important advantages:

- material selection can be operated on the basis of the process specifications;
- membranes are made of low-cost materials;
- they can work at a very high temperature (up to 900°C);
- the compatibility with operating conditions is assured by the fact that every possible porous material can be used as the support.

The main disadvantage is due to the impossibility to produce a membrane with a perfectly determined pore dimension, which can lead to finite selectivity. Anyway, in different experiments, a hydrogen purity of 99% has been obtained, and using more separation stages can virtually afford whichever purityrequired.

1.5.4. Dense Ceramic Membranes

The term "perovskite" identifies a particular mineral with formula $CaTiO_3$ and a large family of metal oxides with a structure similar to the mineral. An ideal perovskite compound have formula ABO_3, where A and B are cations of different dimensions, and a structure in which B (smaller cation) atoms are in the corners of a cube, A (bigger cation) is in the cube center and an oxygen atom is in the center of each edge, leading to the formation of octahedral BO_6 units.

In the early '80s of this past century, it was discovered that some metal oxides with a perovskite structure, such as $SrCeO_3$, could perform proton conduction at high temperature (Iwahara et al., 1981). Since then, a large amount of similar materials have been tested for hydrogen separation from gas mixtures (Roark et al., 2001, Mundschau et al., 2006). It is now recognized that hydrogen cations can dissolve in the perovskite structure thanks to a series of gas-solid reactions that can be schematized by the following equations (Li & Iglesia, 2003)[1]:

[1] Kroger-Vink notation: M_S^C; M = atoms, vacancies (V), interstitials (i), electrons (e), electron holes (h); S = lattice site occupied (atoms, interstitials, etc.); C = no charge (x), positive (•), negative (')

$$1/2O_2 + V_O^{\bullet\bullet} \leftrightarrow O_O^{\times} + 2h^{\bullet} \quad (20)$$

$$H_2 + 2O_O^{\times} \leftrightarrow 2OH_O^{\bullet} + 2e' \quad (21)$$

$$H_2O + V_O^{\bullet\bullet} + O_O^{\times} \leftrightarrow 2OH_O^{\bullet} \quad (22)$$

Once dissolved, protons diffuse by oxygen atoms transferring in the normal grid positions. Thus, hydrogen transport through dense ceramic membranes involves 5 steps: the oxidation of H_2 to H^+ on the feed stream; the adsorption of protons and electrons into the ceramic lattice; the mixed proton-electron conduction; desorption of protons and electrons, and the final reduction of H^+ to H_2 on the permeate side. Usually the process kinetic is determined by the mixed conduction step, whose rate depends on the ambipolar conductivity:

$$\sigma_{amb} = \frac{\sigma_{H^+} \cdot \sigma_{e^-}}{\sigma_{H^+} + \sigma_{e^-}} \quad (23)$$

Where, σ_{H+} and σ_{e-} are respectively the proton and electronic conductivity.

It has been shown that the highest rate of proton conduction is obtained with selectively doped perovskite, such as $SrCeO_3$ e $BaCeO_3$ doped with trivalent cations (Y, Yb e Gd). The general formula of doped perovskites is $A_{1-x}A'_xB_{1-y}B'_yO_{3-\delta}$. Doping B sites with lower valence cations, a high concentration of oxygen vacancies is obtained, that is necessary to have a high proton conductivity. These membranes are used at temperature between 500°C and 1000°C.

The main advantages of dense ceramic membranes are the relatively low cost of membrane materials and the theoretically infinite hydrogen selectivity. On the other hand, these membranes still have several problems, such as thermal stability, acid compounds (H_2O, CO_2, H_2S) resistance; and above all, a permeability that is significantly lower with respect to other classes of membranes, mainly because of the low electronic conductivity of perovskite materials.

1.5.5. Composite Membranes

The hydrogen permeability of dense ceramic membranes is limited by electronic conductivity, which is low for doped perovskites. In order to overcome this limitation, a new approach has been developed since the end of '90s, where a second phase with high electronic conductivity is mixed with the perovskite proton conducting phase. In this way, the composition of each phase can be optimized in order to maximize the respective conductivity (Balachandran et al., 1999, Roark et al., 2001, Roark et al., 2003). If both phases are percolative, the ambipolar conductivity determines hydrogen transport rate, and can be calculated by (Balachandran et al., 1999):

$$\frac{1}{\sigma_{amb}} = \frac{\tau_p}{f_p \sigma_{H^+}} + \frac{\tau_e}{f_e \sigma_{e^-}} \tag{24}$$

Where, f_p and f_e are the volume fraction respectively of the proton conducting (perovskite) phase and the electron conducting phase in the composite materials, while τ_p and τ_e are dimensionless geometric factors. The composition of the membrane can be chosen to obtain the maximum ambipolar conductivity.

The electron conducting phase can be either another ceramic phase or, more frequently, a metallic phase. The best performance are obtained mixing a perovskite phase with a metallic hydrogen conducting phase (palladium or its alloy), also acting as a catalyst for the reaction of hydrogen dissociation on the feed side. Composite membranes are still being developed, and no material seems to be ready for large scale experimentation.

Chapter 2

PALLADIUM MEMBRANES

Palladium membranes have a fundamental role in the research field regarding hydrogen production. The industrial application of such materials could lead to important improvements in the separation of H_2 from hydrogen containing gas streams; and, to the development of innovative applications such as membrane reactors for methane steam reforming and water gas shift reaction. In **Figure 6,** the number of patents and journal articles published per year about hydrogen separation Pd-based membranes is shown.

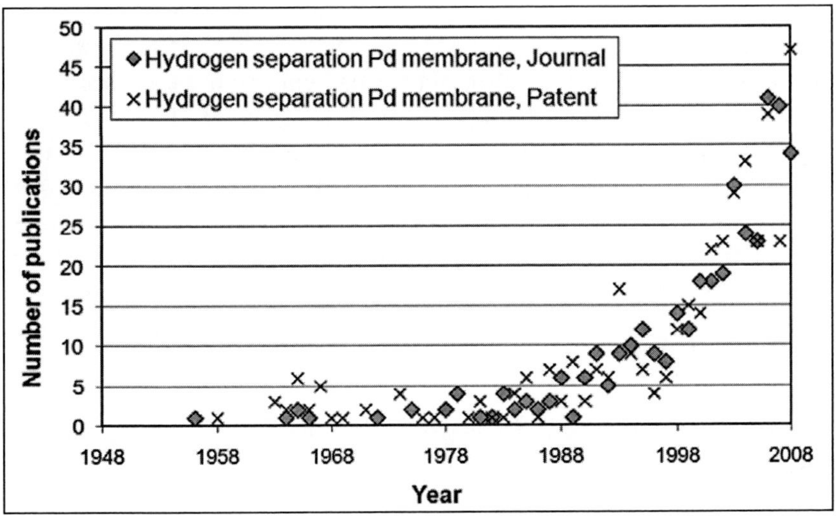

Figure 6. Number of publication per year about hydrogen separation palladium membranes.

Interest in these materials has grown sharply, thanks to some important advantages they have over other described membrane technologies. Palladium membranes are particularly suitable for the separation of hydrogen from syngas after the water gas shift reaction, which is carried out at temperature of 250-400°C (Baade et al., 2001). Moreover they have a really high selectivity coupled with thermal and mechanical resistance. The disadvantages that had limited their large scale application are the low permeability and the high cost, mainly due to the cost of palladium.. Both these problems are strongly reduced by lowering the selective layer thickness. The lower the Pd layer, the higher is the hydrogen flux (see equation 15), and the lower is the precious metal content of the membrane. However mechanical strength has to be preserved. For this reason, attention is focusing on the development of thin (0.1-20 μm) Pd film deposited on porous supports, which are capable to resist to high transmembranal pressure, without increasing flow resistance.

2.1. Pure Palladium and Palladium Alloys Membranes

An important drawback in the use of pure palladium for hydrogen permeation is related to the phase diagram of the system Pd:H (Lewis, 1995, Knapton, 1977, Paglieri & Way, 2002). At temperature lower than 293°C and pressures below 20.15 bar, in the presence of dissolved hydrogen, a β phase forms, which coexists with the α phase which is typical of pure palladium. Both the phases have a face centered cubic structure, but the β phase has a considerably larger unit cell (4.10 Å) respect to the α phase (3.89 Å), so that the phase transition leads to the lattice expansion, with a volume gain from 10% to 20% (Paglieri & Way, 2002). The cyclic expansion and contraction of the palladium membrane when it undergoes H_2 concentration, temperature and pressure variations, can lead to the membrane destruction via the so-called **hydrogen embrittlement**. This problem can be solved by maintaining the membrane above 300°C in the presence of hydrogen (nitrogen purging). This solution cab be however difficult to be applied on large-scale plants.

A more interesting way to solve palladium membranes structural instability is alloying Pd with other metals. The following issues have been studied:

- to suppress the α-β phase transition;

- to obtain the highest possible permeability, maintaining the infinite selectivity;
- to preserve an optimal mechanical strength, even producing thin film membranes;
- to produce less expensive membranes as oppose to pure palladium ones.

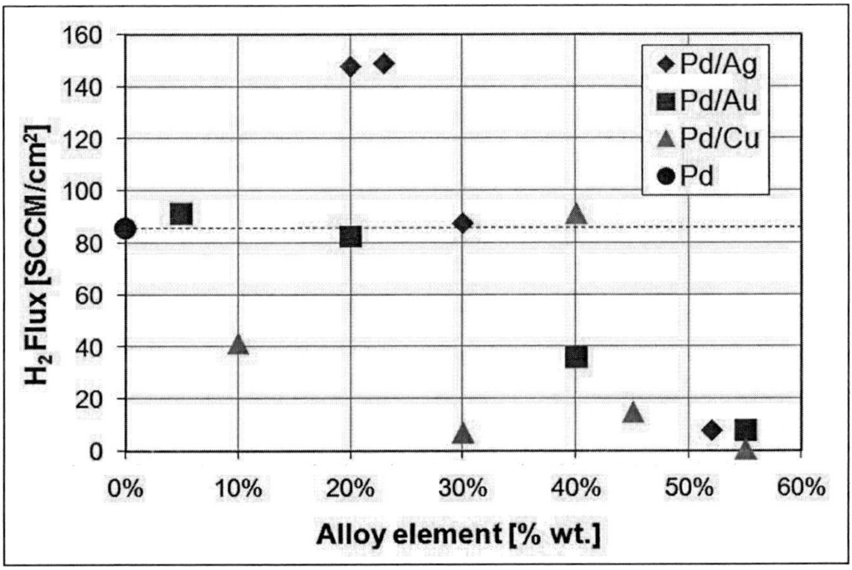

Figure 7. Hydrogen permeation flux at 350°C and 2 bar transmembranal pressure of a 25 μm thick membrane composed by pure Pd or its alloy with Ag, Au, Cu (Knapton, 1977).

Since the '50s, it was shown that adding about 20% of silver to palladium, the desired dimensional stability in presence of hydrogen was obtained (Hunter, 1960). This is due to the fact that adding Ag lowers the critical temperature, under which the miscibility gap is encountered, from 300°C to well below the environmental temperature. At the same time, an important increase of hydrogen permeability in comparison to pure palladium was recorded (**Figure** 7). This increase is maximum for the alloy with 23-25% Agdue to the increased H_2 solubility in the metal lattice (Knapton, 1977). Other metals from the group IB of the periodic table have been tested. Adding gold below 20%, a slight permeability enhancement is found, while flux decreases for further additions (**Figure** 7). The main advantage is the improvement of the chemical resistance to sulphur compounds. Adding copper, permeability is initially reduced, but a maximum is obtained for a 40% Cu composition, whilehigher ratioslower again the flux. The maximum at 40% Cu is due to the formation of an ordered β-phase body centered cubic structure with high hydrogen diffusivity. The solubility, on the other hand, is lowered on alloying with copper, and so the Pd 60%/Cu 40% alloy permeability does not exceed significantly pure palladium permeability (**Figure** 7).

The use of palladium alloys also lead to mechanical strength, dimensional stability, and poisoning resistance (in particular from carbon and sulphuric acid) improvement. Finally, another important issue is the cheapness of Pd/Ag and Pd/Cu alloys with respect to pure palladium. In **Table 5,** the current NIMEX quotation of different metals are reported, while in **Table 6** the calculated cost for square meter of a 15 μm thick membrane of different Pd-based alloys is compared with that of a pure palladium film.

Table 5. Metal NIMEX quotation at the beginning of 2008 and 2009 (www.nymex.com).

Metal	Price (Jan. 2008) [$/kg]	Price (Jan. 2009) [$/kg]
Gold	23000	27000
Palladium	11000	5700
Silver	410	360
Copper	6.5	3.2

Table 6. Material cost of a 15 μm thick film in different alloys, based on metals quotation and average density, and cost variation respect to pure palladium films.

Membrane material	Cost (Jan. 2008) [$/m^2]	Cost (Jan. 2009) [$/kg]
Pure Pd	1300	680
Pd 77%/Ag 23%	1000 (–23%)	520 (–23%)
Pd 60%/Cu 40%	710 (–45%)	370 (–45%)
Pd 95%/Au 5%	1400 (+8%)	800 (+18%)

2.2. Porous Supports

Porous support can be classified as symmetric or asymmetric. A symmetric support has pore dimension and porosity uniform in all the support thickness, while an asymmetric support is constituted by two or more adjacent layers, with thickness and pore dimension decreasing in the direction of the top surface, where Pd has to deposited (**Figure 8**).

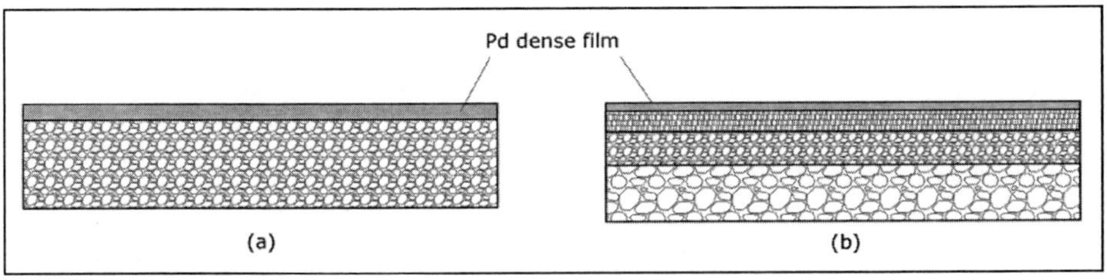

Figure 8. Simplified representation of a palladium film deposited on (a) symmetric and (b) asymmetric porous supports.

Different porous materials have been used in the production of thin-film dense palladium membranes (Paglieri & Way, 2002):

- metallic tubes or discs;
- ceramic tubes, discs or hollow fibers;
- silica glass tubes or discs.

Porous metallic supports are produced by metallic powder sintering. Stainless steel is largely the most used material, but other metals or alloys such as silver and Inconel have been used, too. They are usually symmetric, has non-uniform pore dimension and pores typically bigger than 0.2 μm. Metallic supports have high mechanical strength and can be weld, thus avoiding sealing problems typical of ceramic and glass supports. Moreover they have a coefficient of linear thermal expansion similar to that of palladium, so that stresses caused by thermal cycles are minimized (**Table 7**). One drawback in the use of metallic supports is the difficulty to form very thin (<10 μm) dense palladium films above them, due to the relatively high surface roughness and pores dimension. This problem can be limited to some extent by mechanical treatments of surface abrasion before palladium deposition (Mardilovich et al., 1998). Another disadvantage is the intermetallic diffusion of Pd and Fe into each other at temperature above 400°C. It can be avoided by forming, through different methods, a thin anti-diffusive ceramic layer above the support surface, which blocks the migration of metallic atoms at high temperature (Yepes et al., 2006). Metallic porous tubes or discs are commercially available.

Table 7. Average coefficient of liner thermal expansion between 0°C and 500°C for palladium and different support materials.

Material	α [μm/m°C]
Palladium	12.8
SS AISI 316	17.5
Alumina	7
Vycor glass	0.79

Ceramic supports have a better chemical and thermal resistance with respect to metallic ones, but the thermal expansion coefficient is about half compared to palladium (**Table** 7). They can be symmetric or asymmetric. Symmetric ceramic supports are commercially available and are usually made of α-Al_2O_3. They are characterized by a narrow pore diameter distribution with a mean value of 0.2 μm and a mean porosity of 20%, with high resistance to gas flow. As for metallic supports, the surface roughness is relatively high and it is not easy to obtain defect-free palladium films with thickness below 10 μm. To override this problem, asymmetric ceramic supports have been developed, using 2 to 4 layers. The bottom layer supplies mechanical strength and is the thickest one, usually made of α-Al_2O_3 with a large pore diameter. Through different methods (*dip-coating*, *slip-casting*, *sol-gel*), other layers, increasingly thin and with smaller pore dimension, are deposited above it. These are usually made of γ-Al_2O_3, SiO_2, Zr_2O_3, TiO_2, and have pore diameter between 5 and 50 nm. Above asymmetric supports, thinner and more adherent palladium films are obtained, thanks to the reduced surface roughness of the top layer. Moreover, the flow resistance is lower than for symmetric supports. The only disadvantage is the lower thermal resistance, due to possible sinterization at temperatures between 400°C and 600°C, leading to smaller pores coalescence.

Finally, **silica glass** (Vycor) porous **supports** can be used. They are symmetric, with average pore dimension of 4 nm. They have not been used diffusely because they have low thermal resistance (sinterization above 550°C, or 450°C in presence of moisture) and a thermal expansion coefficient that is extremely lower compared to palladium (**Table** 7).

Considering the cost of the different types of support; we can say that symmetric ceramic or Vycor supports are the cheapest, asymmetric ceramic ones are the most expensive, while metallic ones have an intermediate cost.

2.3. Palladium Film Deposition Techniques

Different techniques have been used to produce palladium-based thin film (0.2-20 μm) membranes supported on porous materials. They are:

- chemical vapor deposition (CVD);
- spray pyrolysis (SP);
- sputter deposition (SD);
- electroplating (EP);
- electroless plating (ELP).

2.3.1. Chemical Vapor Deposition

Chemical vapor deposition (Itoh et al., 2005) is a technique in which both an inorganic salt (e.g. $PdCl_2$) or an organic salt (usually palladium acetate Pd $(CH_3COO)_2$) can be used as palladium precursor: in the latter instance we can speak of *metal organic chemical vapor deposition* (MOCVD). In any case, palladium salt is inserted in a furnace, where the temperature is taken high enough to make the salt sublimate. The gaseous salt is then transferred towards the porous support surface by an inert gas. When it arrives on the surface, its temperature has grown again and the salt decomposes thus depositing metal atoms. The furnace is kept under vacuum in order to eliminate possible contaminants, and the support side opposite to that of deposition is at an even lower pressure, so that gaseous salt flow towards the surface is favored and the dense film formation is possible. Usually organic salts are used because their sublimation and decomposition temperatures are lower (about 180°C and 220 °C respectively for palladium acetate).

The main advantage of CVD is the fact that, for fluid dynamic reasons, palladium flow is directed towards that zone of the support surface where pores still exist. Theoretically, therefore, areas with defects or with metal superabundance are avoided, and very thin film can be obtained (<1 μm). On the other hand, disadvantages are evident: it is not possible to deposit alloy, but only pure palladium; complex equipment is used (high temperature furnace, two vacuum pumps); a lot of energy is consumed. Finally, organic palladium salts are really difficult to synthesize, while inorganic salts require very high decomposition temperatures (>500°C). All these reasons limit the possible large scale application of CVD as a palladium membrane production method.

2.3.2. Spray Pyrolysis

In spray pyrolysis technique palladium precursor is an inorganic salt, usually $Pd(NO_3)_2$. The technique is based on a series of processes: the spraying of a salt aqueous solution, water evaporation, the pyrolysis of the salt at high temperatures, which gives molten metal drops or vaporized metal atoms that finally deposit on the porous support surface. In a typical set up an aqueous solution of palladium nitrate (in case with other alloy element salts) is atomized through an ultrasonic vibrator. The aerosol is then transferred by an oxygen flow to a coaxial burner fed with hydrogen. The decomposition temperature of palladium nitrate is about 700°C, so the temperature is usually maintained above 1000°C. The support, typically tubular, is maintained in the center of the flame and is continuously rotated to achieve a uniform deposition (Li et al., 1993).

Spray pyrolysis requires high fuel consumption and has a low efficiency. Moreover, its control is really difficult, because of the possible metal evaporation in the hotter zones of the support, which can lead to membrane defect formation. For these reasons spray pyrolysis has had no potential for success in recent years.

2.3.3. Sputter Deposition

This technique is classified in the larger category called physical vapor deposition, since it does not involve chemical reactions (McCool et al., 1999). A palladium or palladium alloy metallic target is hit by inert gas (argon) plasma, leading to the detachment of metallic atoms. In the chamber, which is kept under vacuum, metal atoms migrate towards the porous support surface, which is held at high temperature, and they deposit in the form of a thin film. Sputter deposition allows an excellent control of the film composition, by simply changing target composition, and thickness (film as thin as 0.5 μm can be obtained by regulating deposition time). The high cost of equipment for vacuum and plasma generation has however limited its application.

2.3.4. Electroplating

Electroplating is a well-established technique for the deposition of thin metallic film on conductive supports. It is largely used for metals such as nickel and copper. In the case of the production of Pd supported membranes, it involves the semi-reaction of reduction of palladium ions in aqueous solution. This happens at the cathode of the electrolytic cell, which is the metallic porous support itself, negatively charged:

$$Pd^{2+} + 2e^{-} \rightarrow Pd^{0} \quad (25)$$

A platinum anode can be used. Electroplating is often conducted while applying vacuum on the dry side of the support, leading to enhanced film adherence (vacuum electrodeposition) (Nam & Lee, 2000). This technique can be used even for alloy deposition, and allows a good thickness control. It can be applied only with metallic supports.

2.3.5. Electroless Plating

Electroless plating (ELP) is largely the most popular technique for the production of thin-film palladium membranes, as shown in **Figure 9**. The reasons are various: ELP is a simple method, which requires low cost equipment, can be performed on every type of support, and allows the deposition of Pd alloy as well as of pure palladium.

It can be divided in 4 phases:

- support pre-treatment;
- sensitization and activation;
- electroless plating (ELP);
- thermal treatments.

First of all, the porous support, whatever material it is made of, has to be washed accurately in order to remove organic and inorganic contaminants that could then lead to local defects in the membrane. Different cleaning techniques have been used. They all involve a series of ultrasonic bath washings, for instance with hydrogen peroxide, trichloroethylene and 2-propanol (Yeung et al., 1999), or with alkaline and acid solutions (Hou & Hughes, 2003), and acetone (Bosko et al., 2007).

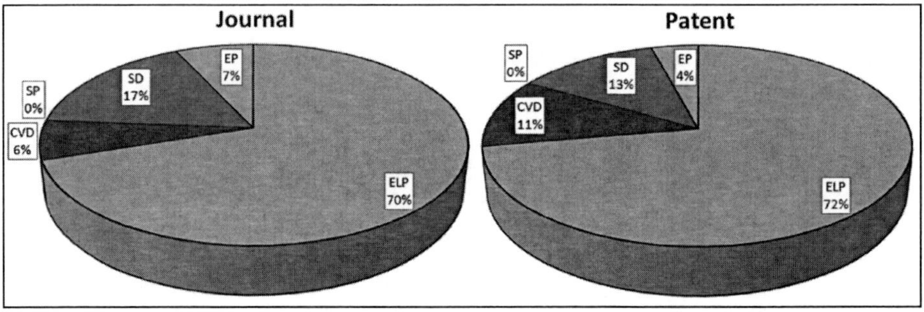

Figure 9. Publications in the period 1998-2008 about hydrogen separation Pd membranes, classified for membrane production method.

Metallic supports are usually treated also in order to form a thin anti-diffusive layer that blocks Pd metal atoms migration into the support at high temperature. Different techniques have been used:

- thin metal oxide or nitride layers can be produced by simply exposing the support surface to oxygen, nitrogen or air at high temperature (Rothenberger et al., 2004);
- a thin silver layer can be deposited by electrodeposition (Tong et al., 2005);
- an α or γ-alumina film can be deposited via dip-coating method (Yepes et al., 2006).

The second step is the surface activation, which is necessary because electroless plating is an autocatalytic reaction, i.e. palladium deposition is catalyzed by palladium itself. Yeung et al. (1999) showed that Pd ELP on a completely polished surface starts after an induction time of about 20 min, while using a surface inseminated with small palladium germs the reaction begins immediately, leading to important advantages such as minor operation costs and higher efficiency. Palladium germs are usually formed on the porous support via the following procedure. Support is first immersed in a Sn^{2+} containing solution, and tin ions are adsorbed on the surface (sensitization). After rinsing with deionized water, the support is immersed in a Pd^{2+} solution (activation). Small palladium grains, with diameter of few nanometers, forms on the support surface through the reaction:

$$Sn^{2+} + Pd^{2+} \rightarrow Sn^{4+} + Pd \qquad (26)$$

The support is then washed to remove tin ions that remain adsorbed on the surface. Dilute acid or basic solutions, or even complexing agents such as EDTA, can be used. The procedure is repeated from 5 to 10 times in order to obtain a uniformly activated support (Yeung et al., 1999). Other activation routes have been tested, such as the deposition of a Pd containing γ-alumina film via sol-gel and slip-casting of a Pd-boehmite solution (Zhao et al., 1998), palladium photocatalytic deposition (Wu et al., 2000) or the deposition of Pd via MOCVD from palladium acetate (Kuraoka et al., 2004).

Electroless plating is a heterogeneous autocatalytic process which takes place in an aqueous solution. It involves oxidation and reduction reactions at the liquid-solid interface, with electron transfer through the interface itself. The plating solution contains a series of components (Djokic, 2002):

- a **metal ion source**, can be whatever metal salt soluble in the plating solution; $PdCl_2$, $Pd(NO_3)_2$ and $Pd(OOCCH_3)_2$ have been used;
- **complexing agents**, whose choice depends on the metal source selected; their function are pH control, salt precipitation prevention, the reduction of the concentration of free metal ions; they are likely to positively influence also the reaction kinetic and the deposit morphology; for palladium deposition ammonium hydroxide and the Na_2EDTA are used;
- a **reducing agent**, whose choice depends on the metal or alloy to be deposited and on reaction conditions; hydrazine (H_2N_4) in alkaline solutions is mainly used for palladium deposition, but also sodium hypophosphite (NaH_2PO_2) has been used, even if with worse results (Cheng & Yeung, 2001).

Using hydrazine, the two semi-reactions of the process are:

$$2Pd(NH_3)_4^{2+} + 4e^- \rightarrow 2Pd^0 + 8NH_3 \qquad [0.0V] \qquad (27)$$

$$N_2H_4 + 4OH^- \rightarrow N_2 + 4H_2O + 4e^- \qquad [1.12V] \qquad (28)$$

And the overall reaction is:

$$2Pd(NH_3)_4^{2+} + N_2H_4 + 4OH^- \rightarrow 2Pd^0 + 8NH_3 + N_2 + 4H_2O \qquad [1.12V] \qquad (29)$$

An alkaline pH of about 10.5 is kept constant by the presence of large amount of non-dissociated ammonia, since it favors reaction thermodynamic. The process begins around palladium germs present on the solid surface:hydrazine and hydroxide anions reacts, producing gaseous nitrogen, water, and electrons, which are transferred to the metallic palladium. Here electrons are used for the reduction of the EDTA stabilized Pd-amine complex, accompanied by gaseous ammonia development and metallic palladium deposition on the pre-existent nuclei. A too high concentration of hydrazine is avoided since it can lead to hydrazine decomposition and to bulk palladium reduction. During the process, a good agitation has to be maintained in order to remove nitrogen and ammonia gas bubbles that tend to adhere to the surface leading to the creation of defects in the membrane.

In order to enhance film adherence, sometimes an additional driving force for the migration of palladium ions towards the surface is furnished. This driving force can be in the form of an absolute pressure in the case of **vacuum ELP**, when vacuum is applied on the opposite side of the support; or, an osmotic pressure, in the case of **osmotic ELP**, when on the opposite side a highly concentrated sucrose or salt solution is fluxed (Paglieri & Way, 2002).

Finally, the produced membrane undergoes a thermal treatment of annealing at temperature between 350°C and 600°C, in nitrogen or hydrogen atmosphere. Yeung et al. (1999) showed that this step leads to a permeability enhancement without selectivity decrease, probably due to the modification of film morphology.

2.3.6. Pd-Alloy Deposition via ELP

Pure palladium is thermally instable (see section 2.1) and is easily poisoned by compounds such as H_2S, CO, CO_2 e H_2O, that are always present in syngas. Alloying Pd with metals such as Cu and Ag phase transition is inhibited and the material turns out to be less expensive, more permeable to H_2 and more resistant to poisons. Typical alloy used for hydrogen separation membranes, whose composition is optimized on permeability basis, are:

- 60% Pd / 40% Cu;
- 77% Pd / 23% Ag.

Electroless plating can be used even for Pd-alloy deposition, through two different methods:

- sequential ELP;
- simultaneous ELP.

In the former method, which is the simpler and more frequently used, palladium is first deposited via classic ELP, then the other metal is deposited through a similar appropriate electroless plating reaction. Procedure can be repeated to improve membrane thickness. Once the two or more metal layers are deposited on the support, a thermal treatment allows the metallic interdiffusion, thus obtaining a film with sufficiently uniform composition (Kikuchi & Uemiya, 1991, Uemiya et al., 1991, Roa et al., 2002, Hou & Hughes, 2003). Both Ag and Cu are usually fed as nitrate. Hydrazine is typically used as the reducing agent for silver ELP, while for copper formaldehyde is preferred.

Pd-Ag alloys have been also deposited via simultaneous co-deposition (Cheng & Yeung, 1999). The reactions of deposition of palladium and silver take place at the same time and in the same plating bath, which contains salts of both the two metals. This technique is less used because it is difficult to control the film composition, due to the different kinetic of metal deposition between Pd and Ag.

2.4. Palladium Membranes Characterization

Palladium-based membranes characteristics (morphology, thickness, grain dimension, composition) are studied through a large series of instruments, including *scanning electron microscopy* (SEM), *energy dispersive X-ray analysis* (EDX), *atomic force microscopy* (AFM), *X-ray diffraction* (XRD), *Auger-electron spectroscopy* (AES), *thermal desorption spectroscopy* (TDS), *X-ray photoelectron spectroscopy* (XPS), *transmission electron microscopy* (TEM), *inductive coupled plasma* (ICP) and *electron-probe microanalysis* (EPMA) (Paglieri & Way, 2002).

Information obtained with this analytical equipment can be used to improve fabrication method in order to obtain better membrane performance.

Film thickness is usually determined by weight gain and by micrograph with SEM or TEM. An average value is usually taken between the two methods, since both methods are to some extent inaccurate. EPMA, used to measure the elemental composition on the cross-section, can as well give indication about film thickness.

Surface morphology is analyzed by SEM, TEM or AFM. XRD is used to determine grain dimension, composition, crystal orientation. XPS and AES allow to detect the presence of contaminants or the segregation of alloy elements. The overall film composition can be determined by dissolution in acid and ICP analysis.

Finally, permeation tests are fundamental to measure membrane performance. Experimental set up is usually composed by a feeding gas system, a permeation cell containing the membrane and inserted in a heated environment, and a system for permeate and retentate analysis.

Chapter 3

EXPERIMENTAL

3.1. METALLIC POROUS SUPPORTS

Porous AISI 316 L stainless steel discs and tubes supplied by Mott® Corporation were used as membrane support. According to Mott®, supports have a 0.1 μm filter grade, which means that a 95% rejection of particle greater than 0.1 μm is guaranteed. According to Mott®, porosity is approximately 15%. Discs have a diameter of 50 mm and a thickness of 1 mm. Tubes have dimensions 6.4 mm o.d., 3.2 mm i.d., with length 100 mm. Each tubular support is welded on both sides to two 100 mm-long tubes of the same material and dimensions, but non-porous, in order to facilitate the membrane high temperature sealing in the permeation cell. Welding is performed directly by Mott®.

3.2. SUPPORTS ABRASION AND CLEANING

Only some tubular supports were abraded before the cleaning and deposition procedure. This treatment was performed in several steps with increasingly finer sandpaper, not only on the porous part but even on the welding zone and the near regions of the dense part. The abraded supports were sonicated several times with deionized water in order to remove steel particles that remain on the surface.

An etching procedure was also tested, that involves the treatment of the abraded surface with an aqueous solution of nitric acid and hydrochloric acid (HNO_3 5 mol/L, HCl 8 mol/L), in order to avoid the complete pores closing.

For all the supports, the following cleaning procedure was used (after the abrasion treatment for abraded supports):

a) sonication in a 0.01 M NaOH solution for 20 min;
b) rinsing and sonication with deionized water for 5 minutes, in order to completely eliminate NaOH;
c) sonication in a 0.01 M HCl solution for 20 min;
d) rinsing and sonication with deionized water for 5 minutes;
e) sonication with acetone for 30 min;
f) drying for 60 min at 120° C.

All sonication steps were performed in an ultrasonic bath at a controlled temperature of 50°C.

3.3. Surface Oxidation

In some experiments, cleaned supports were oxidized in order to form a metal oxide layer which block intermetallic diffusion at high temperature. Heating from environmental temperature to 500°C was performed with a heating rate of 2°C/min. Then the support was hold at 500°C for 12 hr, and finally it was cooled to environment at 2°C/min. Oxidation was performed in a muffle furnace (M110 Heraeus®). Dry support weight was measured via a precision balance (Scaltec® SBC31) before and after oxidation and weight gain was calculated.

3.4. Sensitization and Activation

The composition of the sensitization and of the activation baths is reported in **Table 8**. The equipment is the same used in ELP procedure and is described in section 3.5. Activation was performed only on one side of the disc support, and on the outer surface of the tubular one. Solution was at environmental temperature. The immersion of the support in the stirred sensitization solution was performed for 5 min, followed by thorough rinsing with deionized water and by the immersion in the activation solution for 5 min. After activation, support was cleaned in order to remove adsorbed tin with a 0.01 M HCl solution, and then with deionized water. The entire procedure was repeated 5 times for all membranes.

Each support was then dried at 120°C for 1 h and its weight was measured. For disc supports activation, each step required 50 ml of solution, while for tubes 70 ml were used.

Table 8. Composition of sensitization and activation solutions.

Reagent	Sensitization solution	Activation solution
$SnCl_2 \cdot 2H_2O$ [g/L]	0.1	-
$PdCl_2$ [g/L]	-	0.1
HCl [mol/L]	0.01	0.01

3.5. Electroless Plating

3.5.1. Bath Composition

The composition of the palladium plating bath is reported in **Table 9**. Noteworthy, hydrazine is in stoichiometric amount with respect to palladium, while Na_2EDTA is in large excess. The high ammonia concentration is used because, in this condition, about 96% of dissolved ammonia is non-dissociated (NH_3). Thus, even if ELP reaction consumes OH^- anions and some NH_3 vaporizes, pH is held constant at 10.7 throughout the process by the equilibrium reaction:

$$NH_3 + H_2O \leftrightarrow NH_4^+ + OH^- \tag{30}$$

Electroless plating was conducted at a constant temperature of 60°C. Hydrazine (in 1 M solution) was added in the bath only once it has reached the desired temperature. 50 ml of plating bath were used for disc supports, 70 ml for tubular supports.

Table 9. Composition of palladium ELP bath.

Reagent	Molar concentration [mol/L]
$PdCl_2$	0.02
Na_2EDTA	0.18
NH_4OH	10.4
N_2H_4	0.01

The reaction was stopped after 1 h from the hydrazine introduction, and the membrane was softly rinsed with deionized water. The whole procedure was repeated from 2 to 6 times, depending on the prepared membrane. After that, the membrane was dried at 120°C for 1 h and weighted on a precision balance. The amount of deposited palladium W_{Pd} was estimated by weight gain, and the average film thickness was calculated by the following equation:

$$s = \frac{W_{Pd}}{\rho_{Pd} A} \tag{31}$$

Where, ρ_{Pd} is the metallic palladium density (12023 kg/m^3) and A is the geometric surface area of the support porous part (1963 mm^2 for discs, 1995 mm^2 for tubes). Palladium deposition efficiency, defined as the deposited palladium weight divided by the total amount of palladium in plating solution, was calculated, too.

All reagents used for cleaning, activation and deposition were of analytical grade and was purchased by Carlo Erba®, except for palladium (II) chloride, furnished by Alfa Aesar®.

In **Figure 10** the complete procedure of membrane preparation is schematized.

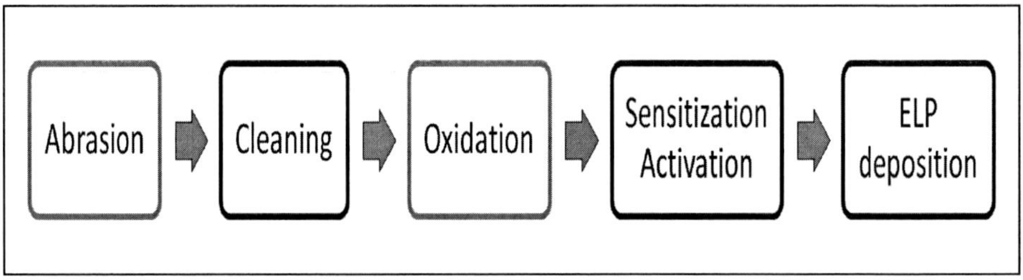

Figure 10. Procedure of membrane preparation. Blue borders identify optional operations.

3.5.2. Disc ELP Equipment

Palladium deposition was effected on only one side of the disc supports. The opposite side was covered with silicon grease before activation was performed. At the end of ELP procedure, grease was accurately removed using a solvent (diethyl ether).

The equipment used for ELP is drawn in **Figure 11**. The disc support was leaned on the bottom of a 60 mm i.d. Pyrex beaker, with the greased side down. 50 ml of the plating solution was poured in the beaker. A mechanical stirrer, with the impeller just above the support top side, is inserted in the beaker in order to improve mass transfer and continuously remove adhering gas bubbles. The beaker itself is put in a water bath heated to 60°C by means of a copper tube coil, in which a solution coming from a thermostatic bath continuously circulates. The heat transfer is improved by the presence of a magnetic stir bar inserted in the water bath, that is directly put on a magnetic stirrer.

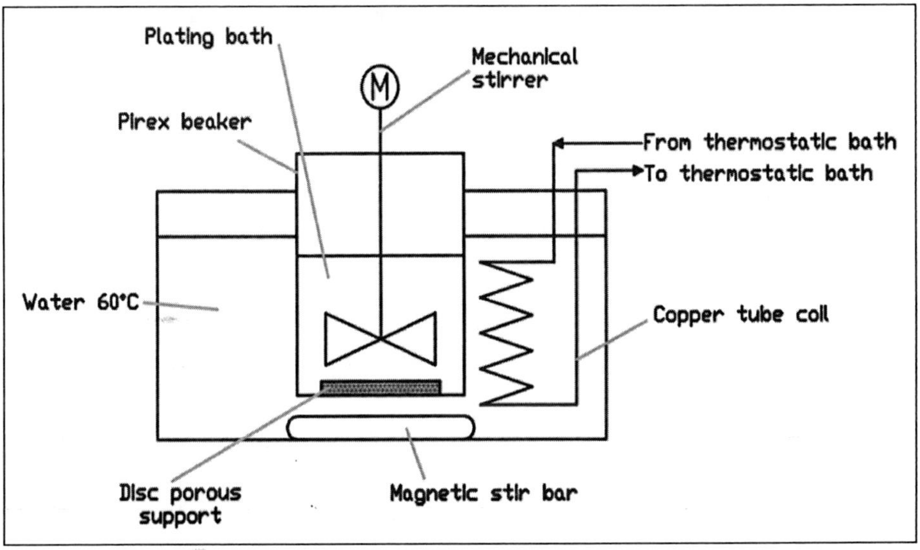

Figure 11. Equipment for palladium ELP on disc supports.

The same configuration was used even for the previous activation procedure, except for the heating system, which is not necessary, since activation was performed at environmental temperature. All operations were conducted under a fume cupboard.

3.5.3. Tube ELP Equipment

Palladium deposition was performed on the outer side of the tube supports. A suitable glass deposition cell was designed and realized (**Figure 12**). The cell is cylindrical, closed on both sides by two rubber-teflon sealing gaskets and two bakelite threaded caps. Caps and gaskets have a central hole, through which the tubular support is inserted so that the porous central part is enclosed in the glass cell, while main part of the two lateral dense welded tubes are outside the cell.

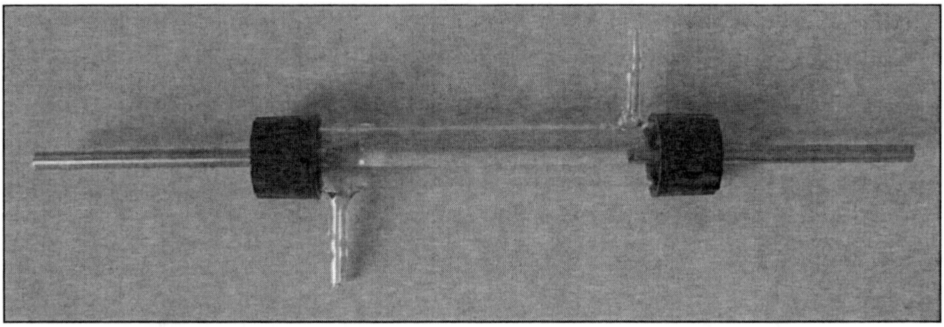

Figure 12. Tubular porous stainless steel support in the glass deposition cell.

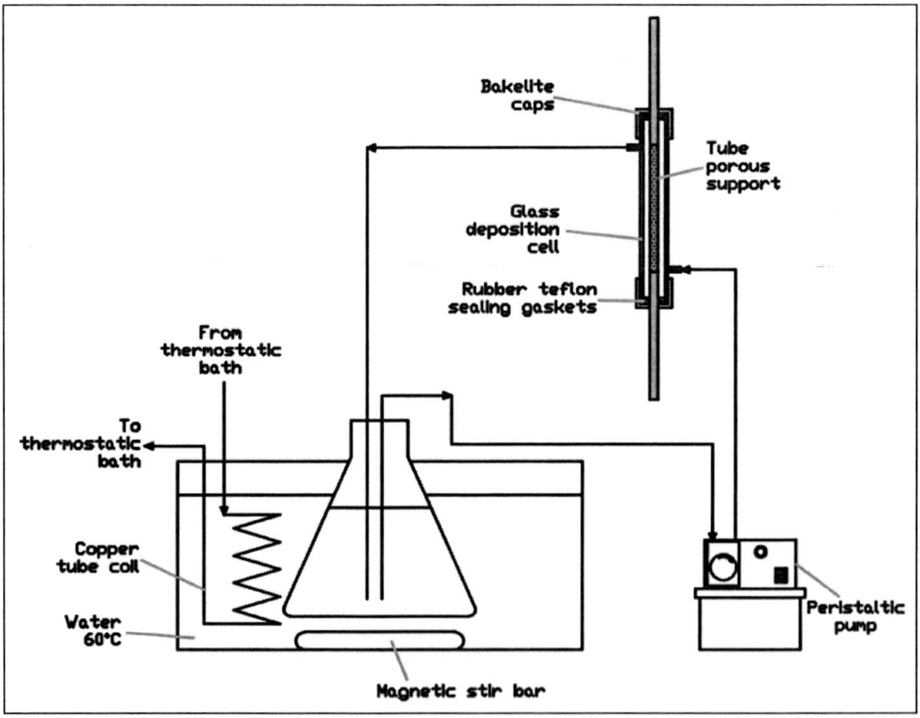

Figure 13. Equipment for palladium ELP on tube supports.

The complete ELP equipment is schematized in **Figure 13**. Plating solution (70 ml) is contained in a conical flask, inserted in a heated water bath, as described for the beaker in section 3.5.2. A PVC tube (7 mm o.d., 4 mm i.d.) is used to transfer the solution from the flask to the bottom of the cell through a peristaltic pump. The solution flows upwards in the cell along the porous part of the tubular support, where ELP reaction takes place, and exits the cell from the top, to go back into the flask. The heated solution is continuously circulated at a high rate given by the pump, so as to avoid cooling and assure good fluid dynamic conditions in the cell, which are required to improve mass transfer and remove gas bubbles. During deposition, the tubular support is axially rotated 180° once every 5 min, while after 30 min of deposition the tube is turned upside down, in order to guarantee deposition conditions as uniform as possible on the whole porous surface. The operation is stopped after 1 h. The same configuration was used for membrane activation.

3.5.4. Vacuum Electroless Plating

Vacuum enhanced electroless plating was experimented in some preliminary trials for tubular supports. The equipment was simply modified by closing with a PVC plug one side of the tubular support and by connecting the other side to a water aspirator linked to a manometer and a variable pressure drop. Absolute pressure in the internal side of the support was maintained at the constant value of 0.9 bar.

This experimentation was early aborted, since in this condition some solution flowed through the pores of the support. This is in agreement with what stated by Mott®, which assigns to 0.1 filter grade AISI 316L media a K_L value of 0.062 bar·hr·m^2/g·Poise·mm, usable in the correlation:

$$\Delta P = K_L \cdot \frac{Q}{A} \cdot \mu \cdot s \tag{32}$$

Where, ΔP [bar] and Q [g/hr] are respectively the pressure drop and the liquid flow rate through the porous media; μ [Poise] is the liquid viscosity, and A [m^2] and s [mm] respectively the surface area and the thickness of the porous media. Assuming that the plating solution viscosity was equal to that of water (0.01 Poise), having an area of 19.9 cm^2 and a thickness of 1.6 mm, for a pressure drop of 0.1 bar, a flow of 0.2 g/hr is obtained. Even if the observed flow is really small, it can lead to undesired deposition of palladium throughout the support thickness. In this way, more palladium is required to form a uniform dense film on the surface. For this reasons, no vacuum was applied during membrane synthesis.

3.6. Morphology Characterization

An optic microscope was used for the morphology characterization of tubular and disc membrane. SEM and EDX analyses on disc fragments (cut before any abrasion, cleaning or deposition operation) were performed using a JEOL 5600 LV instrument.

3.7. Nitrogen Low Pressure Permeation Test

After the tubular membranes were prepared as specified earlier, they were tested for nitrogen permeability. An ideal palladium membrane should not be permeated by gas other than hydrogen. Thus this test is useful to detect the presence of defects in the membrane. The equipment used is drawn in **Figure 14**, and consists of a pure nitrogen gas cylinder connected with the same glass cell used for deposition on tubular supports. The nitrogen pressure is measured by a digital manometer (PI) and is maintained constant by a constant pressure drop (KPD) connected in parallel. Nitrogen can flow only through membrane defects, and exits the cell from the tube inner side. The flow is then measured by a soap bubble flowmeter (FI).

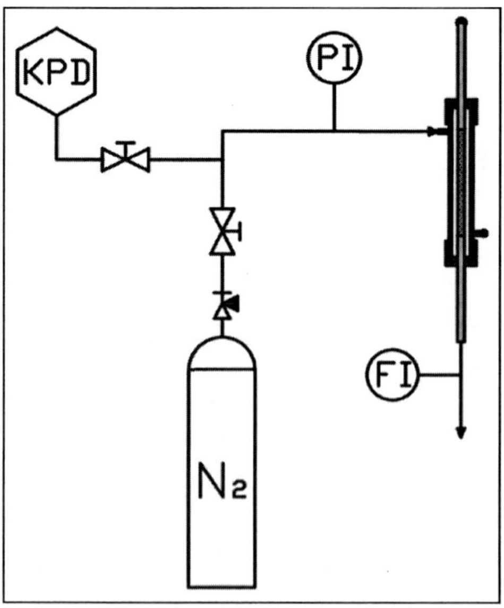

Figure 14. Equipment for nitrogen permeation test.

After that, in order to study the localization of membrane defects, the nitrogen cylinder was connected to the inside of the tubular membrane on one side, while the opposite side was closed. After the gas flow was started, the membrane was inserted in a water filled glass cylinder, and the evolution of gas bubbles was observed.

Chapter 4

RESULT AND DISCUSSION

4.1. POROUS SUPPORTS

SEM analysis was performed both on the surface and the cross-section of the purchased supports (see **Figure 15**). Some surface pores seem to be even larger than 10 μm, while in the cross-sections they all seem to have dimension lower than 5 μm.

Mardilovich et al. (2002) determined the pore size distribution of this material by mercury intrusion, finding out a maximum pore diameter of 4-5 μm, with an average value of 2 μm. However, according to Ayturk et al. (2006), larger pores can be found on the support surface.

Support composition was also analyzed by means of Energy Dispersive X-ray Spectroscopy (EDX). The AISI 316L composition was confirmed.

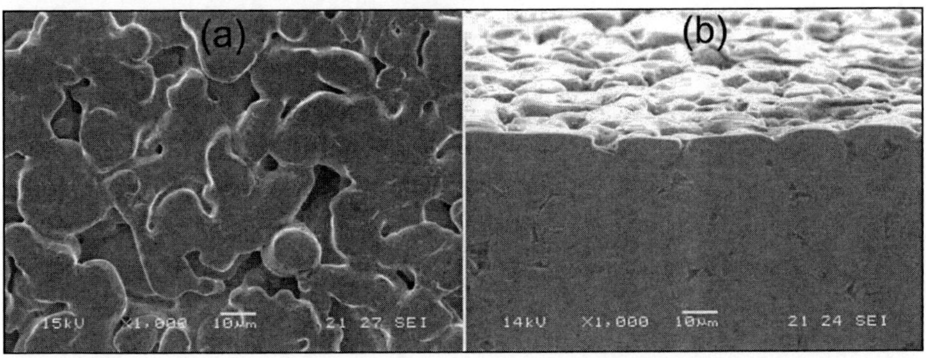

Figure 15. SEM micrographs of porous SS 0.1 μm filter grade support: (a) surface and (b) cross-section.

4.2. Support Abrasion And Etching

Abrasion was performed mainly in order to avoid the formation of local defects near the welding zone of tubular membranes. These defects were detected during nitrogen permeation tests (see Section 4.4).

The abrasion treatment, at the same time, reduces the average dimension of the largest pores nearest to the surface, which are bigger than the underlying region pores, as detected by SEM observations (see Section 4.1). As a consequence, film smoothness after Pd ELP is improved and the minimum film thickness necessary to obtain a defect-free membrane is reduced, in agreement with Mardilovich et al. (2002).

In **Figure 16,** the original support, (a); and the support after abrasion, (b); and after etching with the acid solution, (c) are compared. Noteworthy the abraded support has smaller pores; and above all, a smoother surface with respect to the original support. The etching operation leads pores to open again, even if the morphology is completely different compared to the original support, with finer pores and reduced roughness.

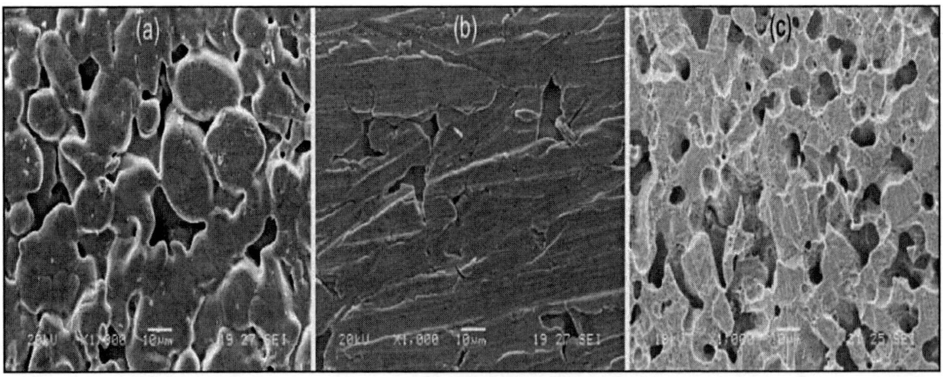

Figure 16. SEM micrographs of porous SS 0.1 μm filter grade support (a) with no treatment, (b) abraded and (c) abraded and etched.

Original and abraded supports were tested for nitrogen permeability in order to evaluate if the abrasion operation could lead to an additional resistance to gas flux during membrane operations. The results are reported in **Figure 17**. Only a slight decrease in nitrogen permeability (about 10%) was originated by the abrasion treatment. Etching operation was therefore avoided.

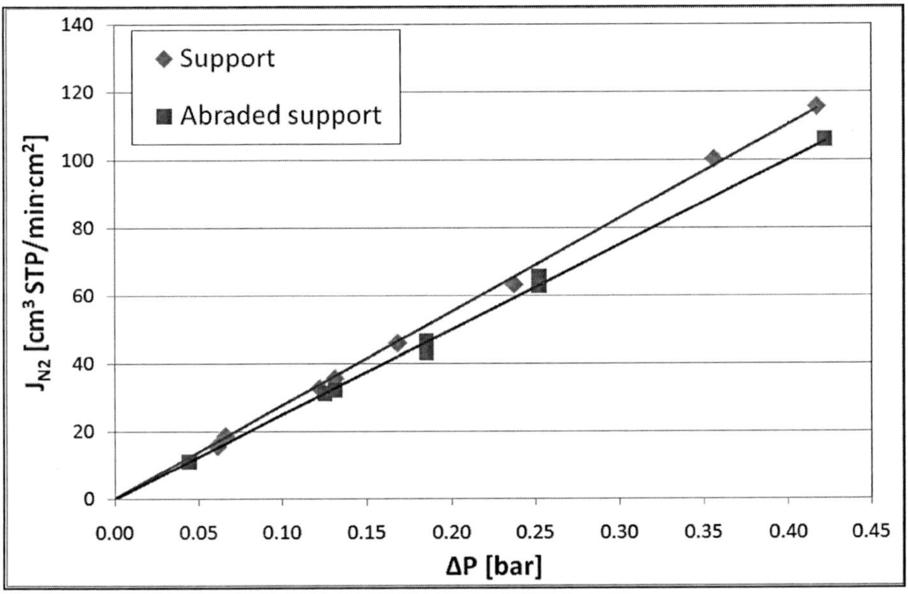

Figure 17. Nitrogen flux Vs transmembrane pressure at 25°C for a tube support (T0) and an abraded tube support (T0, abr.).

4.2. Surface Oxidation

Oxidation experiments were performed on fragments of disc supports, as well as, on integral disc and tubular supports. The typical red color of the surface after the treatment clearly indicates the formation of an oxidized phase. In all experiments, weight improvement was recorded, ranging from 0.05% to 0.07%. The maximum thickness of the oxide layer was calculated by:

- attributing the whole weight gain after oxidation to oxygen atoms that combine with metal atoms;
- assuming that metallic oxides form only on the outer surface of the porous support, supposed equal to the geometric surface;
- allow a mean oxides density of 5.1 g/cm^3 (Ma et al., 2004).

A value of 1.3±0.2 μm was obtained.

Oxidized samples were then observed with SEM. In **Figure 18** the support top layer morphology is shown, before (a); and after (b) oxidation. No evident difference can be noted, other than a less smooth surface for the oxidized support.

Even cross-section was observed, and no detectable film was perceived. This seems to be in contrast with the calculated oxide layer thickness. In fact, the support actual surface is greatly higher than the geometric surface, due to the high surface roughness. Moreover, oxidation is not probably limited to the external side, but it involves even the internal surface of nearest pores. The oxide layer formed during oxidation is therefore very thin, in agreement with what previously found by Ma et al. (2004) for similar temperature oxidation.

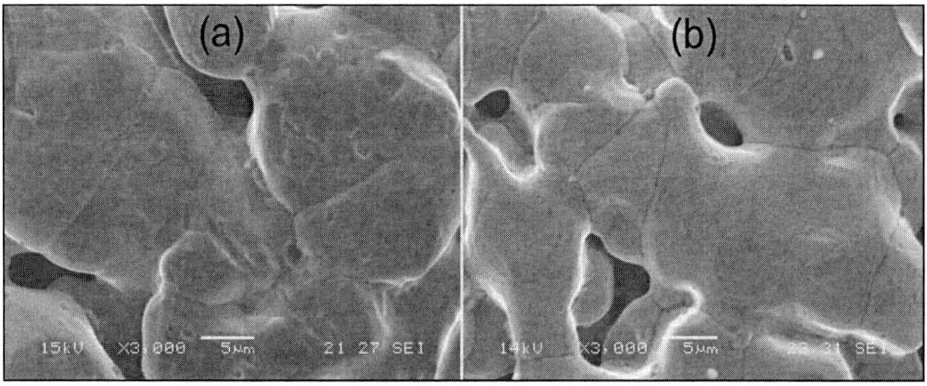

Figure 18. SEM micrographs of porous support top layer (a) after washing and (b) after 500°C air oxidation.

4.3. Palladium Deposition

Four disc membranes, identified with prefix D, and five tubular membranes, identified with prefix T, were prepared. All membranes were pre-oxidized except D2 and T1.

Table 10. Deposited Pd amount, calculated film thickness and deposition efficiency for the prepared membranes.

Geometry (area)	Disc (19.6 cm^2)				Tube (19.9 cm^2)				
ID membrane	D1	D2	D3	D4	T1	T2	T3	T4	T5
Abrasion	No	No	No	No	No	No	Yes	No	Yes
Thermal oxidation	Yes	No	Yes	Yes	No	Yes	Yes	Yes	Yes
ELP cycles	2	4	4	6	4	4	4	6	6
Pd deposited [g]	0.10	0.26	0.28	0.38	0.23	0.26	0.28	0.37	0.40
Pd film thickness [μm]	4.2	10.9	11.7	15.9	9.5	11	11.6	15.5	16.7
Pd deposition efficiency	46%	59%	64%	58%	38%	43%	46%	41%	44%

Membranes T3 and T5 were produced from previously abraded supports.

The number of deposition cycles were varied from 2 (D1) to 4 (D2, D3, T1, T2, T3) and to 6 (D4, T4, T5). In **Table 10**, palladium film thickness and deposition efficiency, calculated on the basis of weight gain, are reported.

In **Figure 19,** the evolution of film thickness for increasing ELP cycles is plotted. Noteworthy the film thickness grows 2.6 μm by average for each deposition. The trend is really similar for discs and tubes, even if a higher amount of plating solution (70 ml) is used for tubes respect to discs (50 ml) at each cycle. The lower palladium deposition efficiency for tubes can be probably explained by the fact that the equipment for deposition on tubes is more complex, and some palladium can deposit inside the PVC tubes, in the flask or on the glass cell walls.

Deposition efficiency grows from 2 to 4 cycles, as a consequence of the improved catalytic effect of the yet deposited palladium, while it slightly decreases going to 6 cycles; probably as a consequence of the reduced adherence of superior Pd grains that can partially detach from the film. The pre-treatment of support abrasion seems to increase the deposition efficiency, as well as the support oxidation.

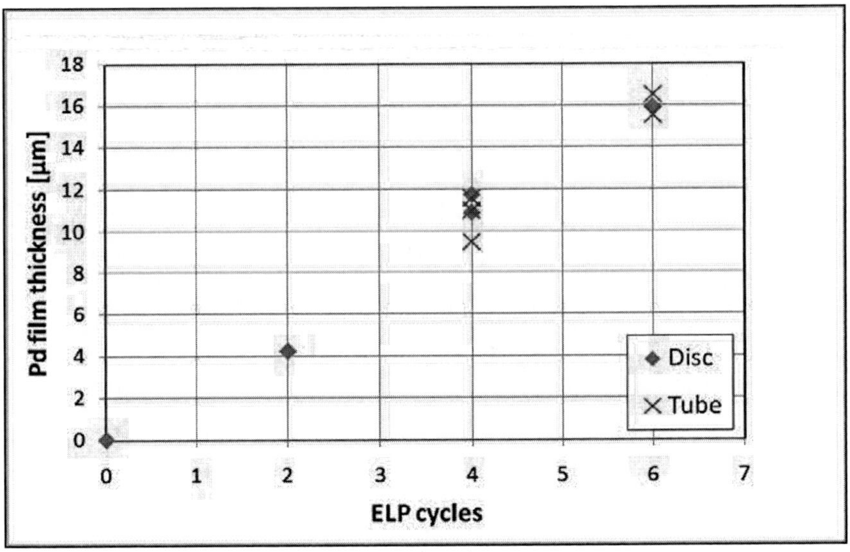

Figure 19. Film thickness as a function of number of deposition cycles

In order to analyze the initial step of electroless plating, D1 membrane surface was observed with SEM and its composition was analyzed via EDX spectroscopy (**Figure 20**). Pd circular grains agglomerates are well discernible on the surface, with grain size of about 1 μm. Large portions of the surface are however not covered by these grains, and most of support pores are open. EDX analysis was performed in different points on the surface, which can be classified as points on the apparently uncovered support surface (point A in **Figure 20**); and, points on deposited palladium grains (point B). In both cases, only palladium was detected by EDX.

First, this confirms that deposited grains are effectively and exclusively made of palladium. In addition, even in those zone where no evident and massive deposition is noticed, a film of palladium thick at least 2 μm is present, since EDX analysis provides the mean composition of a pear-shaped region of matter with a depth of about 2 μm. In fact, observing these regions, it can be noted that small submicronic polygonal palladium grains, really compressed among each other and adherent to the support surface, cover completely the larger stainless steel particles.

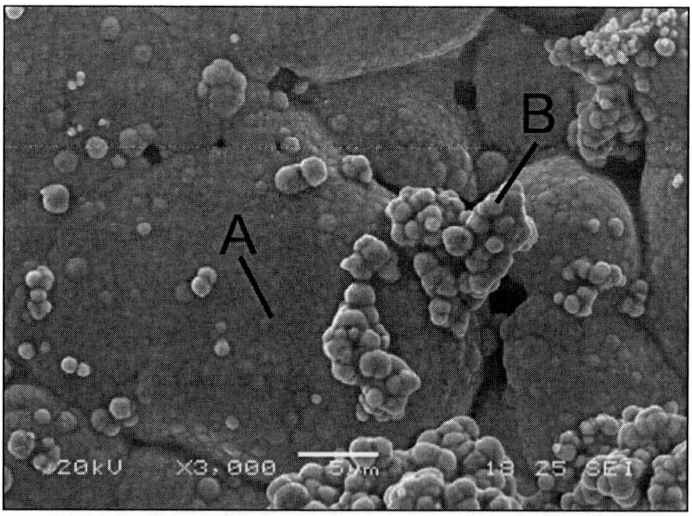

Figure 20. SEM micrograph of D1 membrane top view, with indicated points A and B, where EDX analysis was performed.

SEM analysis was also performed on membranes with 4 and 6 deposition steps. In order to facilitate the operation, two disc supports were cut in three parts with equivalent surface area. Deposition was effected on these fragments (three of which previously oxidized) in the same exact way as for integral membranes. This was made to avoid the cutting operation after electroless plating, necessary for SEM analysis. The cutting can comport morphology modifications and alter results. In **Figure 21,** the evolution of surface morphology for non-oxidized membranes is shown. It can be observed that after 4 ELP cycles, steel support is no more visible, being completely covered by palladium grains, which grows further in dimension after 6 cycles.

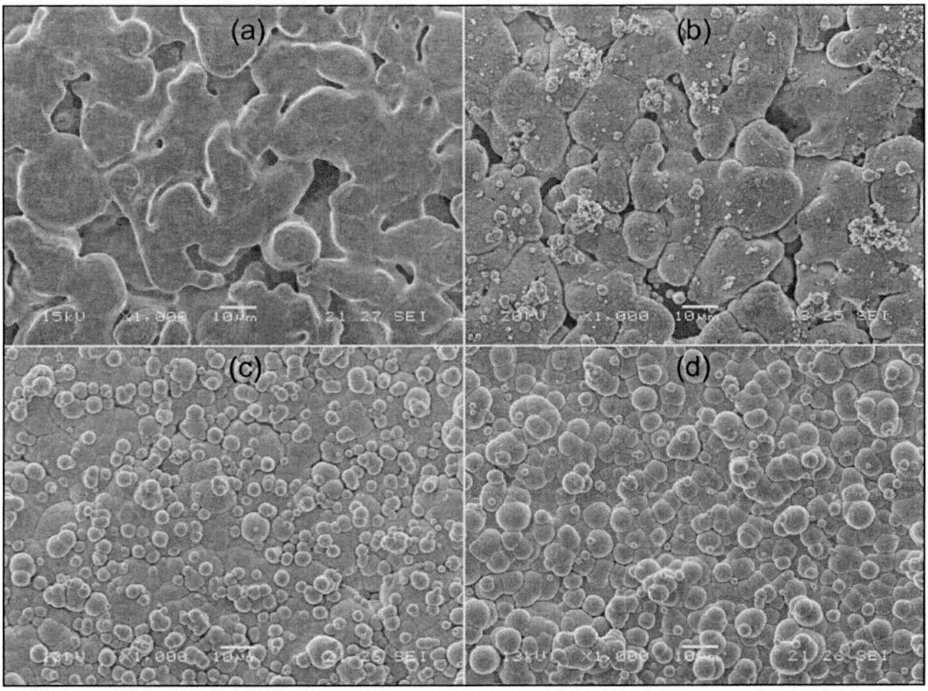

Figure 21. SEM micrographs of cleaned support (a) and non-oxidized membrane top view after 2 (b), 4 (c) and 6 (d) ELP cycles.

In **Figure 22,** the cross sections of the same membranes are reported. After two depositions, the support covering is incomplete, and the mean film thickness varies from 0 to 10 μm, depending mostly on the local characteristics of the underlying support. The mean value, anyway, is in agreement with what is found by weight gain. After 4 cycles, a quite uniform and dense film is formed on all the surface, with a mean thickness slightly above 10 μm, again corresponding to what previously determined. After 6 cycles, thickness increases again, reaching more than 20 μm in some regions. In **Figure 22 (d), it** is particularly evident how palladium penetrates in deepness into the support upper pores, so assuring optimal adherence of the film to the underlying porous structure.

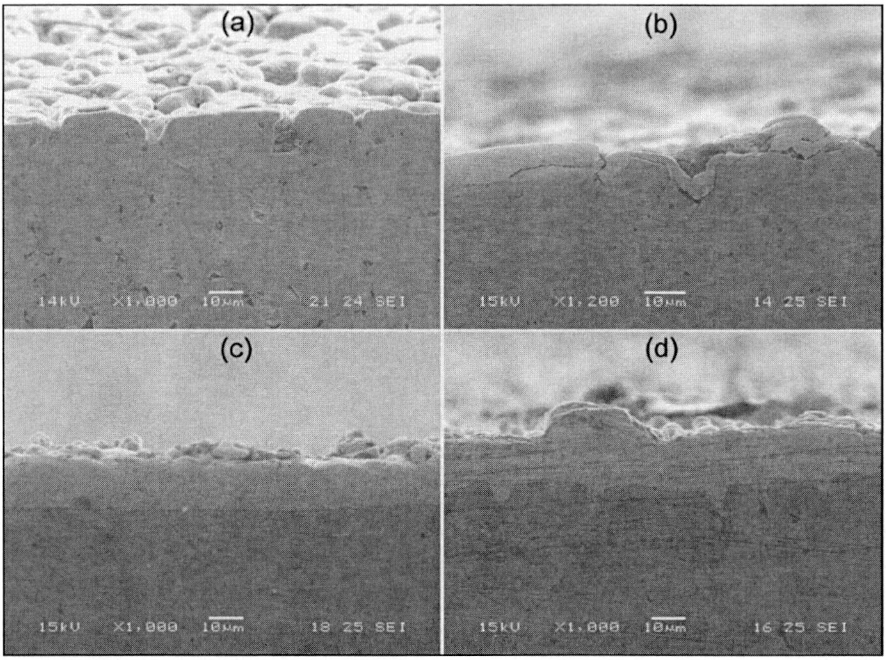

Figure 22. SEM micrographs of cleaned support (a) and non-oxidized membrane cross-section after 2 (b), 4 (c) and 6 (d) ELP cycles.

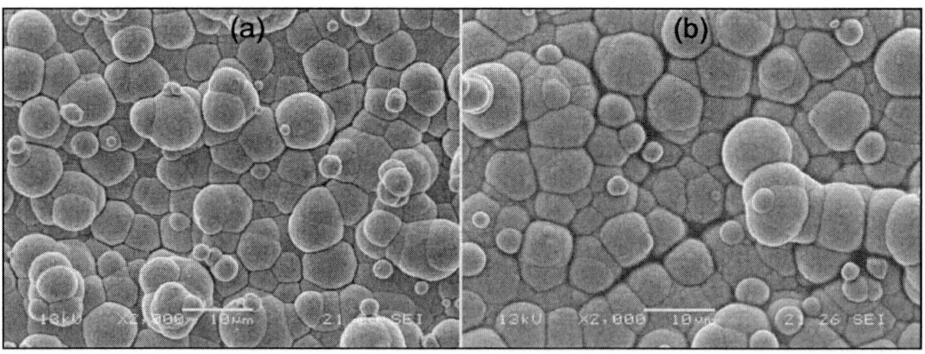

Figure 23. SEM micrographs of a (a) non-oxidized and (b) oxidized membrane top-view after 6 deposition cycles.

Analogous consideration can be made for oxidized membranes. The only important differences are the slightly higher film thickness, retrievable in all steps and in agreement with the results reported in **Table 10**, and the higher mean palladium grain dimension. From **Figure 23** it can be seen that in the non-oxidized membrane after 6 deposition (a), maximum particle dimension is 7 μm, while for the oxidized membrane (b), it can be as high as 10 μm or more.

These observations suggest a mechanism of formation of the palladium film. Palladium first seems to deposit on the steel surface in proximity of the nanometric grains formed in the activation step, and to cover the outer particles of the steel support in the form of small polygonal compressed grains with dimension lower than 1 μm. When all the surfaces are covered, bigger circular grains begin to grow, developing the film thickness. At first, they grow scattered on the surface, depending on the underlying steel surface local morphology. As the palladium amount deposited increases, grains grow in dimension and tend to overlap and pack onto each other, developing a dense film with progressively more uniform thickness. This mechanism of formation guarantees a good adherence between the support and the film, even without applying modified ELP methods such as vacuum ELP.

Finally, pre-oxidation of the steel support seems to improve the palladium deposition, leading to higher deposition efficiency, film thickness and average grain dimension. This can perhaps be attributed to the decreased smoothness of the support surface that can facilitate film formation and adhesion.

4.4. Tubular Membranes Nitrogen Permeability

Results of nitrogen permeation tests at environmental temperature for the tubular membranes are plotted in **Figure 24**, in terms of N_2 flux versus imposed transmembrane pressure.

The nitrogen permeability for tubular supports and for the tubular membranes is calculated from experimental data by using a linear least square approximation. The results are reported in .

Table 11. Film thickness and nitrogen permeability of tubular membranes

Membrane	t [μm]	P_{N2} [cm^3 STP/cm^2 min bar]
T0	0	276.1
T0, abr.	0	249.8
T1	9.5	102.9
T2	11	33.32
T3	11.6	13.48
T4	15.5	5.386
T5	16.7	0.757

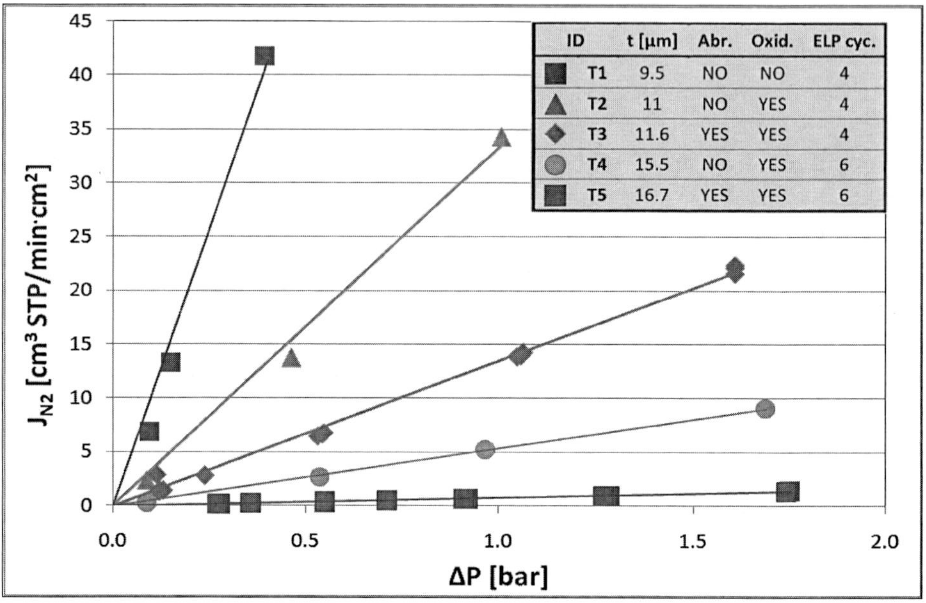

Figure 24. Nitrogen permeability test results: nitrogen flux Vs transmembrane pressure at 25°C.

First, it is worthy to note that no membrane is completely free from defects. Even the thicker membrane, T5, allows a nitrogen flux, normalized with respect to membrane area, of about 0.76 cm^3 STP/min·cm^2 for a pressure difference of 1 bar.

Anyway, at a fixed pressure, nitrogen flux decreases with increasing palladium film thickness, as expected. Noteworthy however the N_2 permeability decrease is not directly proportional to film thickness increase. This is due to the two different pre-treatment of support abrasion and oxidation, which lead to important improvements of membrane tightness.

For instance, for the same number of deposition cycles (4), the pre-oxidized membrane T2 gives a nitrogen flux largely lower (-68%) respect to the non-pre-oxidized one T1, even if the relative increase of film thickness is only of 16%.

About the abrasion treatment, membrane T2 can be compared with T3, and T4 with T5. Permeability reductions respectively of 60% and 86% correspond to film thickness increases of only 5% and 8%.

Tubular membranes T1, T2 and T4 were tested in a water-filled cylinder for defects localization. One result for T4 is shown in **Figure 25**. It can be observed that nitrogen bubbles, for internal overpressure up to 1 bar, forms only in the zone of the membrane near to the welding joint, since membrane defects are concentrated in this area. For membranes T1 and T2, similar results were obtained, even if few defects are present in the central part of these membranes, too, due to the minor film thickness.

Figure 25. Localization of membrane defects through immersion in a water filled graduated cylinder.

Such observation leads to an important conclusion: the discontinuity between the porous part and the dense part of the stainless steel support, in terms of surface roughness, causes a non-perfect formation and adhesion of the palladium film, that tends to detach in that region.

This consideration was confirmed by observing the membrane with an optic microscope. In Figure 26 (a), the interface between the porous part and the welding joint of the support is depicted. A net discontinuity in the surface morphology can be observed. This discontinuity leads to a completely different Pd film aspect in the welding area (Figure 26 (c, d)) with respect to the bulk of the porous support (Figure 26 (b)). In Figure 26 (c, d) several membrane defects can be observed even at limited magnification, which cannot be found in Figure 26 (b). Moreover over the welding joint, the deposited palladium tends to detach from the support.

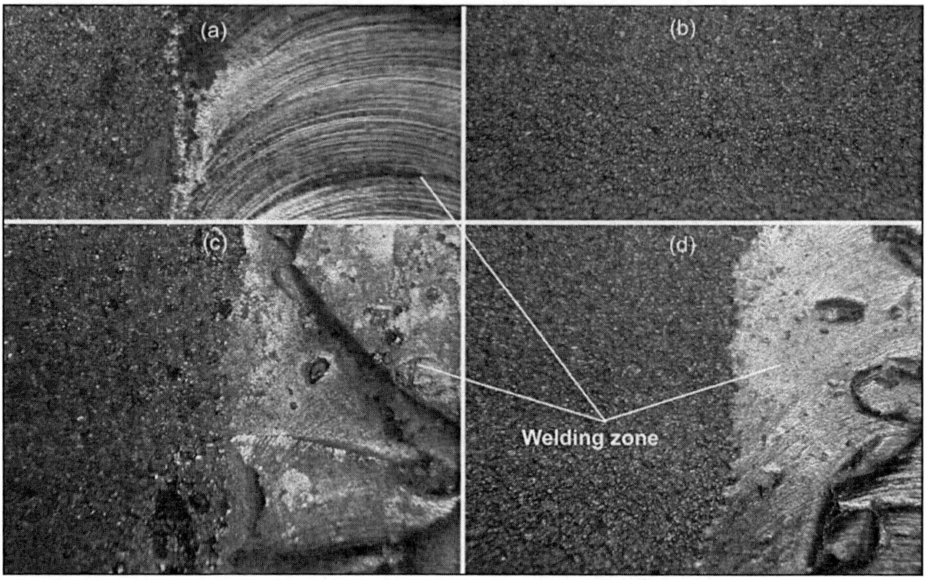

Figure 26. Optic microscope picture of the welding zone in the porous support (a) compared with a Pd membrane after 6 ELP cycles in the porous part (b) and in the welding zone (c, d).

Another possible mechanism for the formation of defects in the membrane near the welding joint could involve modifications of the porous structure induced by the welding operation. These modifications could compromise support quality, for example through pore coalescence. This mechanism was taken into account during this work, but it was considered less probable, since the nitrogen flux measured for the support is even lower than that indicated by Mott® for this kind of porous media with no welding joint. On the contrary, pore coalescence should cause an increase in the flux.

All these considerations led to the employment of the earlier described pre-treatment of support abrasion, in order to limit the formation of film defects. These observations are in agreement with the reported results of nitrogen leakage tests: thanks to the abrasion treatment, the discontinuities between the porous part of the support and the welding joint; and between the joint and the dense part, are leveled. Thus, the film adhesion is improved and the number of defects are strongly reduced. As a consequence, the nitrogen flux reduces considerably.

Chapter 5

PERMEABILITY TEST EXPERIMENTAL SET-UP

5.1. INTRODUCTION

Many publications exist where small area palladium membrane performances are studied, in terms of hydrogen permeability and selectivity with respect to other gases. Even if exceptions exist, the experimental set-ups described in these publications usually have the following limiting characteristics:

- driving force is induced by a sweep gas on the permeate side, or by a limited transmembranal pressure (maximum 5 bar);
- ideal hydrogen selectivity is usually calculated, being the ratio between H_2 and another compound permeabilities, as measured in single-gas permeation tests;
- selectivity is usually found respect to nitrogen only;
- characteristics of poisoning gas resistance or long term flux are not investigated.

In this work, an experimental set-up was designed in order to overcome these limitations, with the precise goal to recreate industrial application conditions of a palladium-based membrane. As a consequence, the following features were outlined:

- operating temperature between room temperature and 550°C;
- feed pressure in the range 1-20 bar;
- opportunity of varying the composition of the feed gas, which can contains H_2, CO_2, N_2, CO, CH_4, H_2O, NH_3, H_2S, so as to simulate a

complex syngas, deriving from different possible fuel thermal conversion processes;

- nitrogen can, if needed, be fed as sweep gas on the permeate side;
- parameters, such as pressure, temperature, gas flow have to be monitored in the key points of the plant.

At the same time, mechanical, sealing and safety issues had to be evaluated. As a consequence, the design of such a plant is really complex. In particular, at high temperature and pressure, the presence of an explosive gas such as hydrogen or of toxic gases such as carbon monoxide and sulphur hydride, has to be carefully considered.

5.2. Experimental Set-Up

In **Figure 27,** the designed experimental set up for permeation tests is schematically represented.

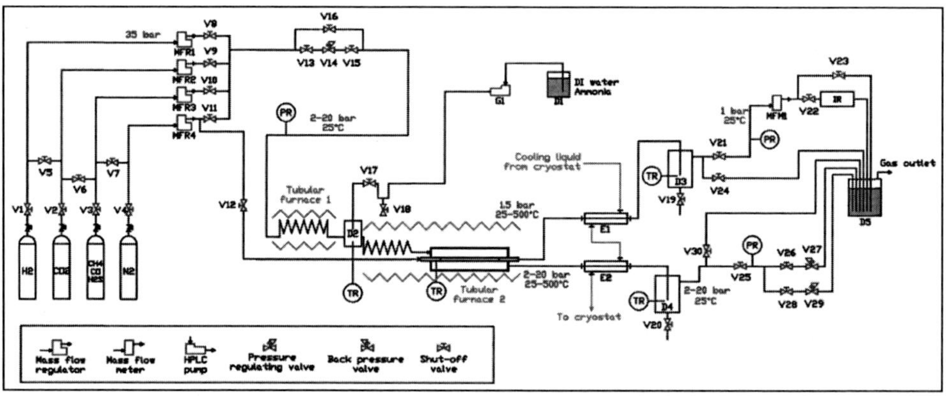

Figure 27. Schematic of permeation test experimental set up.

Four gas cylinders are present, containing the following compounds: hydrogen, carbon dioxide, nitrogen, and finally a mixture with composition 48% CH_4, 48% CO, 4% H_2S by volume. Gas cylinders are connected to the plant through pressure regulators with outlet pressure range 1-50 bar. Each cylinder is then connected to a thermal mass flow regulator. Through them, the desired feed gas composition is obtained, after the different gas lines are collected. The feed pressure can, in case, be reduced by a needle valve (V14), and is monitored by a digital manometer.

Gas mixture is then heated by flowing in a spiral tube inserted in the tube furnace 1. Water or water-ammonia solutions, contained in the vessel D1, are pumped as room temperature liquid into the chamber D2 by the HPLC pump G1. In D2, the gas stream, with a high enough temperature, vaporizes the liquid stream, and the final desired composition is obtained.

At this point, gas enters a second spiral tube that is placed inside the tube furnace 2. Here the desired test temperature is reached, and the feed gas enters the permeation cell (in **Figure 27,** the cell for tubular membranes is schematized, but that for disc membranes can also be used). The temperature inside the cell is accurately controlled by a cascaded control loop, that regulates furnace heating resistances, on the basis of temperature measured by means of thermocouples both in the permeation cell and in the oven chamber.

In the cell, feed gas separates in two streams that are cooled in double pipe heat exchangers (the permeate in E1 and the retentate in E2). In the outer side of the exchanger an aqueous solution of ethylene glycol flows, and is recirculated from a cryostat. After the exchangers, two chambers (D3 and D4),where temperature is measured, assure the separation of condensed water from the gas streams. Two safety lines (those with valves V24 and V30) are present: if gas cooling, for any reason, is not accomplished, downstream equipment can by-passed and saved from thermal shocking.

Both permeate and retentate pressure is monitored by two digital manometers. Retentate pressure is regulated from 2 to 20 bar by means of the back-pressure regulator V27. In parallel with this, a needle valve V29 is present, used to attenuate pressure fluctuations. The permeate stream, on the contrary, is at a pressure slightly above environmental pressure. Permeate hydrogen mass flow is measured by a thermal mass flowmeter, and the presence of other compounds is detected by an FTIR.

Finally, all gas streams are released in atmosphere through the water seal vessel D5. These, in addition to maintain in the permeate line a stable pressure of about 1 atm, is necessary to avoid atmospheric air to enter the plant lines where hydrogen and other flammable gas are present.

As safeguard additional lines are provided that are useful in the initial phase of experiments to eliminate air from all lines by fluxing nitrogen gas; thus avoiding any contact between oxygen and flammable compounds. Moreover, nitrogen can be used as sweep gas in the permeation cell.

5.3. Selected Equipment

The whole equipment, where not otherwise specified, is made of AISI 316L stainless steel.

Considering an area of about 20 cm^2 for the membranes to be tested, and assuming for hydrogen permeation flux a maximum value of 120 cm^3 STP/min·cm^2 (0,89 mol/m^2·s), in agreement with many reviews on palladium membranes, maximum hydrogen flow resulted 2.4 SLPM. Assuming a crude composition 50% H_2 and 50% CO_2, the maximum mixture feed flow is about 5 SLPM. In order to dimension the necessary equipment, two different operating conditions were taken into account:

- C1, at maximum feeding pressure and temperature (20 bar and 550°C); in these conditions hydrogen permeability is maximum, according to equations 15 and 16, and actual volumetric feed gas flow is 0.64 L/min;
- C2, at minimum feeding pressure and maximum temperature (2 bar, 550°C); in these conditions gas density is minimum and actual volumetric gas flow is maximum (1.5 L/min), even if permeability is lower than in C1.

Choosing the optimal gas rate in the tube as 10 m/s, a suitable tube was selected, having 3.2 mm o.d. and 2.2 mm i.d, with 245 bar of maximum pressure rating at 650°C.

Valves were purchased by Swagelok®, as well as all the necessary fittings. Compression fittings were used, with silver plated front ferrules in plant areas where operating temperature is above 200°C. The back-pressure valve selected has 0-34.4 bar pressure regulating range, while the two needle pressure regulating valves have a C_V variable in the range 0.000026-0.004.

Hydrogen, nitrogen and carbon dioxide were supplied by SOL® in 50 L cylinders, containing respectively 10 m^3_{st} of H_2, 10 m^3_{st} of N_2, and 30 kg of liquefied CO_2. The gas mixture is in a 10 L cylinder containing 2 m^3_{st} of gas. Each cylinder is equipped with a two-stage pressure regulator. CO_2 cylinder is equipped with an electric heater to assure liquefied gas vaporization.

Minimum and maximum flow for each gas was calculated on the basis of the range of plant operating conditions (2-20 bar, 150-550°C) and of a variable feed gas composition based on **Table 1**. Mass flow regulators were consequently chosen. Four digital thermal mass flow controllers (model 5850S) were purchased by Brooks Instrument®. Calibration was made by the manufacturer, with the following flow range:

- 0-3 SLPM for H_2;
- 0-3 SLPM for CO_2;
- 0-2 SLPM for N_2;
- 0-0.4 SLPM for the CH_4-CO-H_2S mixture.

The permeate hydrogen flow is measured by a thermal mass flowmeter (model 5860S Brooks Instrument®), having flow range 0.05-3 SLPM.

The tube furnace 1 (Carbolite® MTF 12/25/250) has a heated length of 200 mm. The tube furnace 2 (Nabertherm® RS 170/1000/11), with a heated length of 750 mm and three zone control unit, is equipped with a cascade controller. Water feeding is performed through the HPLC pump PU-2080 by Jasco®, with a flux ranging from 1 μL/min to 10 ml/min and pressure from 30 to 500 bar. K-type thermocouples are used for temperature measurement, while digital manometers Leo Record Ei by Comer® are used for pressure recording.

5.4. Permeation Cells and Other Equipment

Some parts of the plant were completely designed from a fluid dynamic and mechanical point of view and custom-made. These include:

- tube membrane permeation cell;
- disc membrane permeation cell;
- double-pipe heat exchangers (E1, E2);
- evaporating chamber D2;
- two moisture separation vessels D3 and D4.

Mechanical design was made in agreement with European PED normative.

The main issues in the design of permeation cells were:

- the sealing between the membrane element and the inside of the cell, in order to obtain correct values of permeate flux;
- the sealing between the inside of the cell and the external environment, so as to avoid flammable and toxic gas leakage in high temperature areas;
- compact geometry, since the cell has to be inserted in the tube furnace 2 together with the second spiral tube;
- mechanical resistance in the operating conditions.

In **Figure 28,** the sketch of the tubular membrane cell is reported. Note: assured gas leak tightness on both sides is by a conical copper ferrule that is pressed by a screw on the dense part of the tubular support. Swagelok fittings are used to connect the cell and the membrane to the upstream and downstream gas lines.

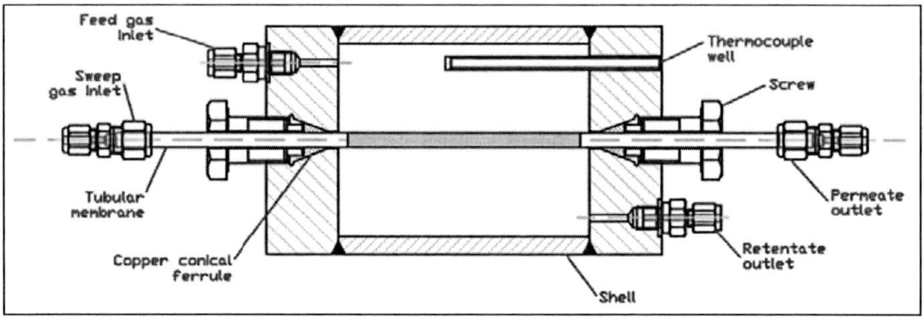

Figure 28. Sectional sketch of the tubular membrane cell.

The disc membrane cell is more complex, and is represented in **Figure 29**. It consists of two flanged connections with copper gaskets. The first, smaller flange acts on the border of the disc membrane and assures the gas tightness between the feed and the permeate side of the cell. The second flange is designed to avoid gas leakage towards the external environment.

In both cells a thermowell is present, in order to allow the measurement of cell temperature by a thermocouple which is protected by the action of corrosive gases.

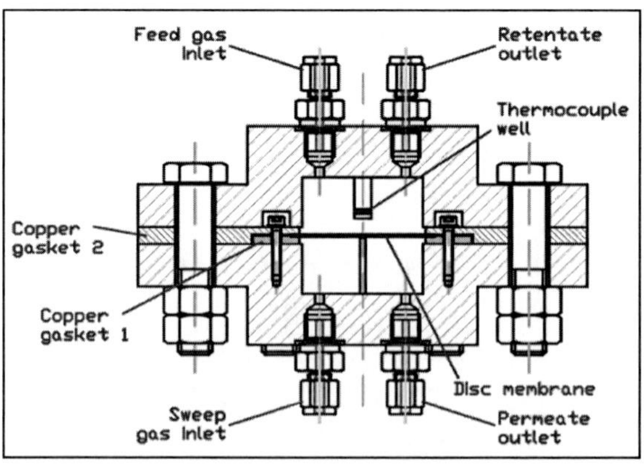

Figure 29. Sectional sketch of the disc membrane cell.

Chapter 6

CONCLUSION

Palladium-based membrane separation is a technology capable of promoting important improvements in the field of hydrogen production from largely available fossil or renewable fuels.

The experimental work conducted reveals important results in the understanding of the process of electroless plating for the production of thin-film palladium membranes, supported above 0.1 μm filter grade porous stainless steel. In particular, effect of an abrasion pre-treatment, of pre-oxidation, of the number of deposition cycles were investigated through weight change determination, optic microscopy, SEM micrographs, EDX analysis, leading to the following main findings:

- palladium film thickness increases of about 2.6 μm for each deposition cycle, in a quasi-linear way;
- the average deposition efficiency is about 60% for disc supports and 40% for tube supports;
- the preliminary oxidation of the support leads to higher deposition efficiency, and to the formation of larger palladium grains and of a thicker film;
- for each prepared membrane, the average film thickness estimated via SEM observations is in good agreement with the thickness calculated by weight gain;
- SEM observation showed that palladium film appears dense and compact above the entire support surface after only 4 deposition cycles, with thickness above 10 μm;

- nitrogen permeability tests demonstrated that the prepared tube membranes, even after 6 deposition cycles, still have some defects, localized near the welding area, due to the imperfect adherence of the film caused by the different surface roughness of the adjacent porous and non-porous zones of the support;
- pre-treatment of surface abrasion (polishing by sandpaper) leads to a smoother support surface; and in turn, to the production of membranes with less defects and a largely minor flux of nitrogen for the same film thickness.

Some modifications to the deposition method used have been suggested by the initial results and will be tested.

- the deposition of palladium alloy with copper and silver, with the technique of sequential electroless plating, will be experimented;
- the heat treatment of annealing of the produced membrane will be studied. It is of fundamental importance, since it can lead to morphology modifications, as well as to membrane performance variations (Yeung et al., 1999).

Disc and tubular membranes will be then tested in the experimental set-up described in section 5. Trials will not be limited to permeability and selectivity measurement:

- a large range of operating conditions will be used, in order to establish possible limitations to the flux laws reported in equations 15 and 16;
- a synthetic syngas with variable composition will be fed to the membranes, testing the consequence on hydrogen flux of the exposure to possibly poisoning gases;
- long term tests, along with morphological analysis with SEM, will indicate possible modifications of the membrane exposed for long period of time to the synthetic syngas;
- structural resistance, as well as sealing effectiveness, will be studied; since these technological problems are important for the industrial application of palladium membrane, as much as film composition or production methods are.

ACKNOWLEDGEMENTS

This work was partially carried out at the Consorzio Polo Tecnologico Magona, Cecina (LI), Italy.

The financial support of MIUR (FISR 2006) is gratefully acknowledged.

The authors want to thank Dr. Piero Narducci, who performed SEM and EDX analysis.

REFERENCES

Adhikari, S., Fernando, S. *Ind. Eng. Chem. Res.* 2006, 45, 875-881.

Asaeda, M., Yamasaki, S. *Sep. Purif. Technol.* 2001, 25, 151-159.

Athayde, A. L., Baker, R. W., Nguyen, P. *J. Membr. Sci.* 1994, 94, 299-311.

Ayturk, M. E., Mardilovich, I. P., Engwall, E. E., Ma, Y. H. J. Membr. Sci. 2006, 285, 385-394.

Baade W. F.; Parekh U. N.; Raman V. S. In *Kirk-Othmer Encyclopedia of Chemical Technologies*; Kirk-Othmer, R. E.; 4th Ed.; John Wiley & Sons: New York City, NY, 2001; Vol. 13, pp. 759-808.

Baker R. W. *Ind. Eng. Chem. Res.* 2002, 41, 1393-1411.

Balachandran, U., Lee, T. H., Dorris, S. E. (1999). Development of mixed-condicting ceramic membrane for hydrogen separation. *http://www.osti.gov/bridge/servlets/purl/12434-GLrtdd/webviewable/*

Bosko, M. L., Yepes, A., Irusta, S., Eloy, P., Ruiz, P., Lombardo, E. A., Cornaglia, L. M. *J. Membr. Sci.* 2007, 306, 56-65.

Breck, D. W. *Zeolite Molecular Sieves*; John Wiley & Sons: New York City, NY, 1973; p. 636.

Cheng, Y. S., Yeung, K. L. *J. Membr. Sci.* 2001, 182, 195-203.

Cheng, Y. S., Yeung, K. L. *J. Membr. Sci.* 1999, 158, 127-141.

Collot, A. G. *Prospect for Hydrogen from Coal*; IEA Clean Coal Centre: London, UK, 2003.

De Lange, R. S. A., Keizer, K., Burggraaf, A. J. *J. Membr. Sci.* 1995, 104 , 81-100.

Djokic, S. S. *Mod. Asp. Electrochem.* 2002, 35, 51-133.

Freeman, B., Pinnau, I. *Trends Polym. Sci.* 1997, 5(5), 167-173.

Gillette, J. L., Kolpa, R. L. (2007). Overview of interstate hydrogen pipeline systems. *http://www.osti.gov/bridge*

Graham, T. *Phil. Trans. Roy. Soc.* 1866, 156, 399.

Gregory, D. P., Pangborn, J. B. *Ann. Rev. Energy.* 1976, 1, 279-310.

Grimmer, P. G., Xiaobing, X., Evenson, C. R., Wright, H. A., Brereton, C., Wolfs, W. (2006). An analysis of dense hydrogen membranes as a means of producing a CO_2 rich stream consistent with the CO_2 capture requirements of a FuturGen plant. *http://www.eltronresearch.com/docs/Hydrogen_Membrane_for_CO2_capture .pdf*

Haussinger, P., Lohmuller, R., Watson, A. M. In *Ullmann's Encyclopedia of Industrial Chemistry*, 6th Ed.; Wiley-VCH: Weinheim, DE, 2003; Vol. 17, pp 85-240.

Hou, K., Hughes, R. *J. Membr. Sci.* 2003, 214, 43-55.

Hunter, J. B. *Platinum Met. Rev.* 1960, 4(4), 130-131.

Itoh, N., Akiha, T., Sato, T. *Catal. Today* 2005, 104, 231-237.

Iwahara, H., Esaka, T., Uchida, H., Maeda, N. *Solid State Ionics* 1981, 3/4, 359-363.

Kikuchi, E., Uemiya, S. *Gas Sep. Purif.* 1991, 5, 261-266.

Knapton, A. G. *Platinum Met. Rev.* 1977, 21(2), 44-50.

Koros, W. J., Ma, Y. H., Shimidzu, T. *J. Membr. Sci.* 1996, 120, 149-159.

Kuraoka, K., Zhao, H., Yazawa, T. *J. Mater. Sci.* 2004, 39, 1879-1881.

Lewis, F. A. *Int. J. Hydrogen Energy* 1995, 20 (7), 587-592.

Li, L., Iglesia, E. *Chem. Eng. Sci.* 2003, 58 , 1977-1988.

Li, Z. Y., Maeda, H., Kusakabe, K., Morooka, S., Anzaib, H., Akiyamab, S. *J. Membr. Sci.* 1993, 78, 247-254.

Lin, H., Van Wagner, E., Freeman, B.D., Toy, L.G., Gupta, R.P. *Science* 2006, 311, 639-642.

Ma, Y. H., Akis, B. C., Ayturk, M. E., Guazzone, F., Engwall, E. E., Mardilovich, I. P. *Ind. Eng. Chem. Res.* 2004, 43, 2936-2945.

Mardilovich, I. P., Engwall, E. E., Ma. , Y. H. *Desalination* 2002, 144, 85-89.

Mardilovich, P. P., She, Y., Ma, Y. H., Rei, M. H. *AIChE J.* 1998, 44, 310.

Matteucci, S., Yampolskii, Y., Freeman, B. D., Pinnau, I. In *Materials Science of Membranes for Gas and Vapor Separation*; John Wiley & Sons: New York City, NY, 2006; Chapter 1, pp. 1-49.

McCool, B., Xomeritakis, G., Lin, Y.S. *J. Membr. Sci.* 1999, 161, 67-76.

Meteyer, M. *Cryogas Int.* 2008; February.

Mundschau, M. V., Xie, X., Evenson, C. R., Sammells, A. F. *Catal. Today* 2006, 118 , 12-23.

Nam, S. E., Lee, K. H. *J. Membr. Sci.* 2000, 170, 91–99.

Nomura, M., Ono, K., Gopalakrishnan, S., Sugawara, T., Nakao, S.I., *J. Membr. Sci.* 2005, 251 , 151-158.

Paglieri, S. N., Way, J. D. *Sep. Purif. Methods* 2002, 31(1), 1-169.

Pandey, P., Chauhan, R. S., *Prog. Polym. Sci.* 2001, 26, 853-893.

Prabhu, A. K., Oyama, S. T. *J. Membr. Sci.* 2000, 176 , 233-248.

Roa, F., Block, M. J., Way, J. D. *Desalination* 2002, 147, 411-416.

Roark S. E., Mackay R., Sammells A. F. (2003). Hydrogen Separation Membranes for Vision 21 Energy Plants. *http://www.netl.doe.gov/technologies/coalpower/gasification*

Roark, S. E., Mackay, R., Sammells, A.F. (2001). Hydrogen separation membranes for vision 21 fossil fuel plants. *http://www.osti.gov/bridge/servlets/purl/832722-y6KzsP/native/832722.pdf*

Robeson, L. M. *J. Membr. Sci.* 1991, 62, 165-185.

Rothenberger, K. S., Cugini, A. V., Howard, B. H., Killmeyer, R. P., Ciocco, M. V., Morreale, B. D., Enick R. M, Bustamante., F., Mardilovich I. P., Ma Y. H. *J. Membr. Sci.* 2004, 244, 55-68.

Shaole, L. J., Berry, D. A., Syamlal, M. In *Kirk-Othmer Encyclopedia of Chemical Technologies*; Kirk-Othmer, R. E.; 4th Ed.; John Wiley & Sons: New York City, NY, 2001; Vol. 6, pp 771-832.

Sholl D.S., Ma Y.H. *Mater. Res. Soc. Bull.* 2006, 31, 770-773,

Sing, K. S. W., Everett, D. H., Haul, R. A., Moscou, L., Pierotti, R. A., Rouquerol, J., Siemineniewska, T. *Pure Appl. Chem.* 1985, 57, 603-619.

Stephan, K., Dahm, B. In *Hydrogen Energy Progress XI,* Veziroglu, T. N.; Proceedings of the 11th World Hydrogen Energy Conference, 11th; International Association for Hydrogen Energy, Coral Gables: Stuttgart, DE, 1996, Vol. 2, pp. 1483–1492.

Tong, H. D., Gielens, F.C., Gardeniers, J. G. E., Jansen, H. V., Van Rijn, C. J. M., Elwenspoek, M. C., Nijdam, W. *Ind. Eng. Chem. Res.* 2004, 43, 4182-4187.

Tong, J., Kashima, Y., Shirai, R., Suda, H., Matsumura, Y. *Ind. Eng. Chem. Res.* 2005, 44, 8025-8032.

US DOE (Department of Energy) (2002), Energy Efficiency and Renewable Energy, www.eere.energy.gov, Vehicle and Fuel Cells, Fact of the week #205, Feb. 2002.

Uemiya, S., Sato, N., Ando, H., Kude, Y., Matsuda, T., Kikuchi, E. *J. Mater. Sci.* 1991, 56, 303-313.

Uhlhorn, R. J. R., Keizer, K., Burggraaf, A. J. *J. Membr. Sci.* 1992, 66 , 271-287.

Ward, T.L., Dao, T., *J. Membr. Sci.* 1999, 153, 211-231.

Wipf, H. *Phys. Scr.* 2001, T94, 43-51.

Wu, L. Q., Xu, N., Shi, J. *Ind. Eng. Chem. Res.* 2000, 39, 342-348.

Yepes, D., Cornaglia, L.M. , Irusta, S., Lombardo, E.A. *J. Membr. Sci.* 2006, 274, 92-101.

Yeung, K. L., Christiansen, S. C., Varma, A. *J. Membr. Sci.* 1999, 159, 107-122.

Yon, C. M., Sherman, J. D. In *Kirk-Othmer Encyclopedia of Chemical Technologies*, Kirk-Othmer, R. E.; 5th Ed.; John Wiley & Sons: New York City, NY, 2003; Vol. 1, pp 617-663.

Zhao, H. B., Pflanz, K., Gu, J. H., Li, A. W., Stroh, N., Brunner, H., Xiong, G. X. *J. Membr. Sci.* 1998, 142, 147-157.

INDEX

A

acetate, 13, 31, 35
acetone, 33, 40
acid, 22, 28, 33, 35, 38, 39, 50
activation, 17, 20, 33, 34, 35, 40, 41, 42, 43, 45, 57
adhesion, 20, 57, 59, 60
adiabatic, 9
adsorption, ix, 8, 9, 13, 15, 16, 19, 20, 22
aerosol, 32
AFM, 37
Ag, 14, 17, 27, 28, 36, 37
agent, 35, 37
air, 6, 34, 52, 63
alkaline, 33, 35, 36
alloys, 17, 28, 29, 37
ammonia, 4, 14, 36, 41, 63
ammonium hydroxide, 35
annealing, 36, 68
application, 12, 14, 26, 31, 32
aqueous solution, 32, 33, 35, 39, 63
Arrhenius law, 20
atmosphere, 36, 63
atmospheric pressure, 9, 10
atomic force microscopy (AFM), 37
atoms, 21, 22, 29, 31, 32, 34, 51
automotive application, 2

B

back, 5, 18, 45, 63, 64
biomass, 3, 7, 8
blocks, 29, 34
bubbles, 36, 42, 45, 47, 58
buildings, 3

C

capillary, 3, 18, 19
carbon, 1, 2, 3, 5, 6, 7, 8, 14, 28, 62, 64
carbon dioxide, 1, 2, 7, 14, 62, 64
carbon monoxide, 5, 6, 62
carbon nanotubes, 3
carrier, 1, 2, 3
casting, 30, 35
catalyst, 6, 23
cell, 1, 2, 5, 26, 33, 38, 39, 43, 44, 46, 53, 63, 65, 66
ceramic, ix, 12, 17, 20, 22, 23, 29, 30, 71
ceramic microporous membranes, 12
CH4, 5, 8, 13, 19, 61, 62, 65
chemical reactions, 7, 32
chemical vapor deposition, 21, 31
CHP, 2
Chrysler, 5
cleaning, 7, 33, 39, 42, 46

CO2, 5, 6, 7, 8, 9, 10, 14, 19, 22, 36, 61, 64, 65, 72
coal, 2, 5, 6, 7
combustion, 1, 2, 5, 6
compatibility, 21
components, 5, 8, 9, 19, 35
composition, 5, 8, 23, 27, 32, 36, 37, 40, 41, 49, 53, 54, 61, 62, 63, 64, 68
compounds, 9, 13, 17, 22, 27, 36, 62, 63
concentration, 16, 18, 22, 26, 35, 36, 41
condensation, 18, 19
conduction, 21, 22
conductivity, 22, 23
configuration, 43, 45
consumption, 3, 4, 32
contaminants, 31, 33, 37
control, 32, 33, 35, 37, 63, 65
conversion, ix, 1, 3, 5, 10, 62
cooling, 45, 63
copper, 17, 27, 32, 37, 43, 66, 68
costs, 4, 5, 34
covering, 55
cracking, 17
critical temperature, 13, 27
cryogenic, 3
CVD, 31
cycles, ix, 17, 29, 52, 53, 55, 56, 58, 60, 67, 68

D

decomposition, 31, 32, 36
decomposition temperature, 31, 32
defect formation, 32
defects, ix, 31, 33, 36, 46, 50, 58, 59, 60, 68
density, 3, 28, 42, 51, 64
Department of Energy, 73
deposition, ix, 17, 21, 29, 31, 32, 33, 34, 35, 36, 37, 39, 42, 43, 44, 45, 46, 52, 53, 54, 56, 57, 58, 67, 68
desorption, 9, 13, 15, 22, 37
diffusion, 12, 13, 15, 16, 18, 19, 20, 29, 40
diffusivity, 13, 20, 28
discs, 29, 39, 42, 53
dissociation, 16, 23
distribution, 3, 4, 5, 19, 30, 49
doped, 22, 23

E

electrodeposition, 33, 34
electron, 21, 22, 23, 35, 37
electron microscopy, ix, 37
electron-probe microanalysis, 37
electrons, 21, 22, 36
energy, ix, 1, 2, 3, 6, 9, 10, 31, 37, 73
energy density, 3
Energy Efficiency and Renewable Energy, 73
engines, 2, 5
environment, 38, 40
equilibrium, 16, 41
etching, 39, 50
ethylene, 3, 63
ethylene glycol, 63
Europe, 3
evaporation, 32
evolution, ix, 47, 53, 55
external environment, 65, 66

F

fabrication, 37
feeding, 6, 9, 10, 11, 12, 38, 64, 65
fiber, 12, 14
film, ix, 20, 26, 27, 28, 29, 31, 32, 33, 34, 35, 36, 37, 38, 42, 45, 50, 52, 53, 54, 55, 56, 57, 58, 59, 60, 67, 68
film formation, 31, 57
film thickness, ix, 37, 42, 50, 52, 53, 55, 56, 57, 58, 59, 67, 68
films, ix, 17, 20, 28, 29, 30
financial support, 69
flow, 6, 16, 17, 18, 26, 30, 31, 32, 45, 46, 47, 62, 63, 64, 65
fluid, x, 31, 45, 65
formaldehyde, 37
fossil, ix, 1, 2, 3, 5, 8, 10, 67, 73
FTIR, 63
fuel, ix, 1, 2, 4, 5, 6, 7, 17, 32, 62, 73

G

gas, ix, 1, 3, 4, 5, 6, 7, 9, 10, 11, 12, 13, 14, 15, 16, 18, 19, 20, 21, 25, 26, 30, 31, 32, 36, 38, 42, 45, 46, 47, 50, 61, 62, 63, 64, 65, 66
gas separation, 11
gases, 7, 11, 12, 13, 14, 18, 20, 61, 66, 68
gasification, ix, 2, 3, 6, 7, 8, 73
generation, 2, 5, 32
glass, 3, 21, 29, 30, 43, 44, 46, 47, 53
grains, 34, 53, 54, 55, 57, 67
greenhouse gas, 7, 10

H

H_2, 1, v, 1, 2, 3, 4, 5, 6, 8, 10, 11, 13, 14, 15, 16, 19, 20, 22, 25, 26, 27, 36, 61, 64, 65
heat, 2, 6, 9, 43, 63, 65, 68
heat transfer, 43
heating, 5, 40, 43, 63
heterogeneous, 35
high pressure, 9, 14
high temperature, 16, 21, 29, 31, 32, 34, 39, 40, 62, 65
Honda, 5
HPLC, 63, 65
hydrazine, 35, 36, 41
hydro, 6, 17
hydrocarbons, 6, 17
hydrochloric acid, 39
hydrogen, ix, 1, 2, 3, 4, 5, 6, 7, 9, 10, 12, 13, 14, 15, 16, 17, 18, 19, 20, 21, 22, 23, 25, 26, 27, 32, 33, 34, 36, 46, 61, 62, 63, 64, 65, 67, 68, 71, 72
Hyundai, 5

I

IB, 27
IEA, 71
mmersion, 40, 59
industrial, ix, 3, 25, 61, 68
industrial application, 3, 25, 61, 68
industry, 4, 17
inert, 12, 31, 32
infinite, 17, 19, 22, 27
inorganic, 31, 32, 33
interface, 16, 35, 59
internal combustion, 2, 5
ions, 33, 34, 35, 36
Italy, 69

K

KL, 45

L

large-scale, 26
lattice, 15, 16, 17, 21, 22, 26, 27
law, 15, 16, 20
leakage, 60, 65, 66
limitations, 61, 68
linear, 29, 57, 67
localization, ix, 47, 58
London, 71
low molecular weight, 20

M

mass transfer, 42, 45
measurement, 65, 66, 68
mechanical treatments, 29
membranes, ix, 9, 10, 11, 12, 13, 14, 17, 18, 19, 20, 21, 22, 23, 25, 26, 27, 29, 31, 32, 33, 34, 36, 37, 40, 46, 50, 52, 54, 55, 56, 57, 58, 63, 64, 67, 68, 72, 73
metal organic chemical vapor deposition, 31
metal oxide, 21, 34, 40
metals, 14, 26, 27, 28, 29, 32, 36, 37
methane, 2, 5, 25
methanol, 4, 14
microscope, 46, 59, 60
microscopy, ix, 37, 67
migration, 29, 34, 36
missions, 1, 2
MOCVD, 31, 35

modules, 10, 12, 14
moisture, 7, 30, 65
moisture content, 7
molecules, 12, 13, 18, 19, 20
morphological, x, 68
morphology, ix, 17, 35, 36, 37, 46, 50, 51, 54, 57, 59, 68
MSC, 21
MTBE, 3

N

nanometers, 34
nanotubes, 3
natural, 3, 4, 5, 6, 7
natural gas, 3, 4, 5, 6, 7
Nb, 14
New York, 71, 72, 73, 74
nitrate, 32, 37
nitrogen, ix, 4, 7, 17, 26, 34, 36, 46, 47, 50, 57, 58, 60, 61, 62, 63, 64, 68
nitrogen gas, 46, 63

O

observations, 50, 57, 60, 67
organic, 7, 31, 33
osmotic pressure, 36
oxidation, ix, 3, 6, 22, 35, 40, 51, 53, 58, 67
oxide, 20, 34, 40, 51, 52
oxygen, 6, 21, 22, 32, 34, 51, 63

P

palladium, ix, 14, 17, 23, 25, 26, 27, 28, 29, 30, 31, 32, 33, 34, 35, 36, 37, 41, 42, 43, 44, 45, 46, 53, 54, 55, 56, 57, 58, 59, 61, 64, 67, 68
particles, 39, 54, 57
permeability, ix, x, 11, 13, 14, 16, 17, 19, 20, 22, 23, 26, 27, 36, 46, 50, 57, 58, 61, 64, 68
permeation, x, 11, 14, 15, 18, 26, 27, 38, 39, 46, 50, 57, 61, 62, 63, 64, 65
perovskite, 21, 22, 23
petrochemical, ix, 3, 9, 13
pH, 35, 36, 41
phase diagram, 26
plants, 2, 14, 26, 73
plasma, 32, 37
platinum, 33
PMA, 37
poison, 17
poisoning, 17, 28, 61, 68
polarization, 18
polymeric membranes, 12, 13
pore, 18, 19, 20, 21, 28, 29, 30, 49, 60
porosity, 19, 28, 30, 39
porous, ix, 14, 17, 18, 19, 21, 26, 29, 30, 31, 32, 33, 34, 39, 42, 43, 44, 45, 49, 50, 51, 52, 55, 59, 60, 67, 68
pressure, ix, 6, 8, 9, 10, 11, 12, 13, 14, 16, 18, 20, 26, 27, 31, 36, 45, 46, 51, 57, 58, 61, 62, 63, 64, 65
prevention, 35
producers, 17
production, ix, 1, 2, 3, 4, 5, 7, 8, 9, 14, 17, 20, 25, 29, 31, 32, 33, 34, 67, 68
protons, 22
PSA, ix, 8, 9, 10, 11
purification, 9, 11, 21
PVC, 44, 45, 53
pyrolysis, 31, 32

Q

quasi-linear, 67

R

Raman, 71
range, 7, 9, 11, 20, 61, 62, 64, 65, 68
reagents, 6, 42
recovery, 9, 10, 11, 13
refineries, 3, 4, 8, 9
regulators, 62, 64
resistance, 12, 14, 17, 22, 26, 27, 28, 30, 50, 61, 65, 68
room temperature, 61, 63
roughness, 29, 30, 50, 52, 59, 68

S

safeguard, 63
safety, x, 3, 62, 63
salt, 31, 32, 35, 36
scanning electron microscopy, ix, 37
selectivity, 9, 11, 13, 14, 17, 18, 19, 20, 21, 22, 26, 27, 36, 61, 68
SEM, ix, 37, 46, 49, 50, 51, 52, 53, 54, 55, 56, 67, 68, 69
sensitization, 33, 34, 40, 41
separation, ix, 1, 9, 10, 11, 12, 14, 17, 18, 19, 20, 21, 25, 26, 34, 36, 63, 65, 67, 71, 73
series, ix, 1, 9, 14, 18, 20, 21, 32, 33, 35, 37
silica, 20, 29, 30
silica glass, 29, 30
silver, 17, 27, 29, 34, 37, 64, 68
sintering, 29
SiO2, 20, 30
smoothness, 50, 57
sol-gel, 21, 30, 35
solubility, 13, 14, 27
species, 6, 11, 19
spectroscopy, 37, 53
stability, 14, 17, 22, 27, 28
stainless steel, ix, 21, 39, 44, 54, 59, 64, 67
steel, ix, 4, 21, 29, 39, 44, 54, 57, 59, 64, 67
streams, 10, 25, 63
strength, 14, 17, 20, 26, 27, 28, 29, 30
sulphur, 4, 17, 27, 62
surface area, 14, 42, 45, 54
surface diffusion, 18, 19, 20
surface roughness, 29, 30, 52, 59, 68
synthesis, ix, 1, 4, 5, 45

T

temperature, 2, 6, 11, 12, 13, 14, 16, 17, 18, 20, 21, 22, 26, 27, 29, 31, 32, 34, 36, 39, 40, 41, 43, 52, 57, 61, 62, 63, 64, 65, 66
thermal expansion, 29, 30
thermal stability, 14, 17, 22
thermal treatment, 33, 36, 37
thin film, 27, 31, 32
tin, 34, 35, 40
TiO_2, 20, 30
toxic, 62, 65
Toyota, 5
transfer, 1, 35, 43, 44
transition, 15, 17, 26, 36
transmembrane, 14, 51, 57, 58
transmission electron microscopy, 37
transport, 3, 5, 12, 15, 18, 19, 20, 22, 23
trichloroethylene, 33
tubular, 12, 17, 32, 39, 40, 41, 43, 45, 46, 47, 50, 51, 52, 57, 63, 66, 68

U

uniform, ix, 28, 29, 32, 37, 45, 55, 57

V

vacancies, 21, 22
vacuum, 9, 31, 32, 33, 36, 45, 57
values, ix, 16, 18, 65
vapor, 21, 31, 32
viscosity, 45
Volkswagen, 5

W

water, 3, 6, 7, 25, 26, 32, 34, 36, 39, 40, 41, 43, 44, 45, 47, 58, 59, 63
water gas shift reaction, 6, 7, 25, 26
weight gain, ix, 37, 40, 42, 51, 53, 55, 67
welding, 39, 50, 58, 59, 60, 68

X

XPS, 37
X-ray diffraction (XRD), 37
X-ray photoelectron spectroscopy (XPS), 37

Y

yield, 6, 7